164 Topics in Current Chemistry

Organic Peroxygen Chemistry

Editor: W. A. Herrmann

With contributions by
W. Adam, K. Dear, J. Fossey, L. Hadjiarapoglou,
H. Heaney, E. Höft, D. Lefort, M. Rüsch,
R. Sheldon, M. Sojka, J. Sorba, S. Warwel

With 43 Figures and 23 Tables

Springer-Verlag
Berlin Heidelberg GmbH

This series presents critical reviews of the present position and future trends in modern chemical research. It is addressed to all research and industrial chemists who wish to keep abreast of advances in their subject.

As a rule, contributions are specially commissioned. The editors and publishers will, however, always be pleased to receive suggestions and supplementary information. Papers are accepted for "Topics in Current Chemistry" in English.

ISBN 978-3-662-14931-7   ISBN 978-3-540-47494-4 (eBook)
DOI 10.1007/978-3-540-47494-4

Library of Congress Catalog Card Number 74-644622

Typesetting: Th. Müntzer, Bad Langensalza;

51/3020-5 4 3 2 1 0 — Printed on acid-free paper

Volume Editor

Prof. Dr. *Wolfgang A. Herrmann*
Anorganisch-Chemisches Institut
der Technischen Universität München,
Lichtenbergstraße 4, 8046 Garching, FRG

Editorial Board

Preface

Oxygen is the most abundant element on earth, and oxidation processes are important ones in industry. Oxygen constitutes approximately 25% by weight of the atmosphere, 45% of the lithosphere, and more than 84% of the hydrosphere of our planet, be it in elementary form or as chemical compounds. However, the oxygenation potential of elementary oxygen is limited, mostly for kinetic reasons. So more efficient forms of natural occurence and of man-made nature became important (Fig. 1): ozone (O_3), hydrogen peroxide (H_2O_2), and *tert*-butyl hydroperoxide ($^tC_4H_9OOH$).

In spite of oxygenation reactions being so important for the production of both bulk and fine chemicals, there is a striking lack of research on this topic in industry and university laboratories. If we include catalytic chemistry using metals and metal compounds, then the impact of oxidation chemistry can be demonstrated by figures: Along homogeneous catalytic routes, approximately fourteen million tons per year of organic products are being made as opposed to "only" seven million tons of hydroformylation products and no more than 1.5 million tons of products from homogeneous catalytic hydrogenation. As far as organometallic catalysis is concerned, the underestimation of oxidation processes has led to a lack of knowledge in the field of high oxidation-state organometallics.

It is interesting to see how little cognitive progress has been made in metal-catalyzed oxidation chemistry: The majority of present-days methodology originates from the earlier decades of this century, with the names of Harries, Criegee, Rieche and others symbolizing major breakthroughs in organic peroxgen chemistry.

The present volume of "Topics in Current Chemistry" is focused on the exciting and useful chemistry of "peroxidic oxygen", cf. organic peroxy reagents such as peroxides, percarboxylic acids and their derivatives. In this context, the chemical efficiency of hydrogen peroxide and its *inorganic* congeners has to be born in mind once again. For this very reason, ORGANIC PEROXYGEN CHEMISTRY (ORPEC) has gained the interest of organic and inorganic chemists dealing with stoichiometric as well as catalytic oxygenation reactions. Some contributions that were highlights at the ORPEC conference held at the Technische Universität München on March 29 and 30, 1991, have been arranged as review articles and will be presented in the present monograph. It is to be hoped that a stimulating effect will be experienced by many readers.

That we have to master oxygen and peroxygen chemistry better in the future than we have done previously, is also the result of a change that is beginning to

Fig. 1. The discoverers of oxygenation reagents: *Scheele* (oxygen), *Schönbein* (ozone), and *Thénard* (hydrogen peroxide).

take place with regard to oxidation feedstock chemicals: As chlorine and chlorine oxides begin to lose their importance as industrial oxidation and bleaching chemicals, the "cleaner" alternatives such as hydrogen peroxide are coming to the fore. This development can already be seen in the paper bleaching process technology. An oxidizing agent that is acceptable in industry has to be cheap, efficient, easy to handle, and selective in its reactivity. Not very many chemicals meet this requirement but O_2, O_3, H_2O_2 and organic derivatives are among them. It should be emphasized here that the full potential of the latter is not yet fully known, e.g. the extraordinarily high selectivity shown by the *dioxiranes* as reviewed on pp. 45–62 by W. Adam et al. The editor is quite confident that the present book will remind us of the broad scope of the applications of "peroxidic oxygen".

Elemental *oxygen* was first isolated by C. W. Scheele in Sweden and, independently, by J. Priestley in England (1773/74). They both used the decomposition of metal oxides, nitrates, and carbonates to isolate pure "dephlogistated air". The element was named "oxygen" by A. L. Lavoisier in 1777 who erroneously thought this element to be an essential constituent of all acids (oxygen means "to form an acid"). The first industrial production of liquid oxygen by the Linde process started in 1896.

Ozone as the more reactive elemental modification of oxygen was discovered by Ch. F. Schönbein in 1840 who named it after the intense odour. The generation of ozone in reasonable amounts was made possible in 1857 by Werner von Siemens who constructed an electrical discharge apparatus. The discovery and development of the "ozonolysis reagents" followed in 1903 by C. B. Harries who applied it to the selective oxidation of olefins.

Hydrogen peroxide was discovered in 1818 by L.-J. Thénard, who treated barium peroxide (BaO_2) with sulfuric acid and thus obtained a relatively dilute solution of hydrogen peroxide. The first industrial production of hydrogen peroxide on the basis of this reaction originates from the German companies Schering (Berlin) and Merck (Darmstadt). The process was practiced until the year 1966 by Bariumoxyd GmbH, a subsidary of Kali-Chemie, in Bad Honningen (Germany) for the reason that the baryte ($BaSO_4$) formed as the side product was particularly useful. Up to the mid-sixties, the so-called "Weißenstein process" — anodic oxidation of H_2SO_4 and hydrolysis of the resulting $H_2S_2O_8$ — was used before the present-day "Anthroquinone process" went on stream. This latter reaction — basically a simple hydrogen-plus-oxygen reaction — was invented by the IG Farben chemists Riedl and Pfleiderer during World-War II. The scale-up from a successfully operating pilot plant to a large-volume production in Silesia did not come into operation because of the War. Production then started in 1955 in Memphis/USA.

<table>
<tr><td>Technische Universität München
January, 1993</td><td>Wolfgang A. Herrmann</td></tr>
</table>

Source of pictures: G. Bugge (ed.), Das Buch der Großen Chemiker, vol 1, p 288a, 464a, p 368a, Reprint of Verlag Chemie 1929. Reproduced with permission of the Verlagsgesellschaft Chemie mbH, Weinheim (Germany)

Attention all "Topics in Current Chemistry" readers:

A file with the complete volume indexes Vols. 22 (1972) through 163 (1992) in delimited ASCII format is available for downloading at no charge from the Springer EARN mailbox. Delimited ASCII format can be imported into most databanks.

The file has been compressed using the popular shareware program "PKZIP" (Trademark of PKware Inc., PKZIP is available from most BBS and shareware distributors).

This file is distributed without any expressed or implied warranty.

To receive this file send an e-mail message to:
SVSERV@DHDSPRI6.BITNET.
The message must be: "GET /CHEMISTRY/TCC_CONT.ZIP".

SVSERV is an automatic data distribution system. It responds to your message. The following commands are available:

HELP	returns a detailed instruction set for the use of SVSERV,
DIR *(name)*	returns a list of files available in the directory "name",
INDEX *(name)*	same as "DIR",
CD ‹name›	changes to directory "name",
SEND‹filename›	invokes a message with the file "filename",
GET ‹filename›	same as "SEND".

Table of Contents

Novel Organic Peroxygen Reagents
for Use in Organic Synthesis

Harry Heaney

Department of Chemistry, Loughborough University of Technology, Leicestershire, LE11 3TU, U.K.

Table of Contents

In this review, I will discuss recent work involving new safe peroxidic reagents that have been developed in order to overcome some of the hazards associated with reagents that have been used in the past but which are now no longer in favour. I will also comment on other work using the reagents that my coworkers and I have developed and also, where appropriate, draw comparisons with other peroxygen reagents, especially *m*-chloroperoxybenzoic acid (MCPBA).

Topics in Current Chemistry, Vol. 164
© Springer-Verlag Berlin Heidelberg 1993

1 Introduction

A number of oxidations that have been reported in the past have used high concentration (>85%) aqueous hydrogen peroxide in order to generate, for example, trifluoroperoxyacetic acid. However, perhaps the most widely reported peroxidic oxidising agent in the past twenty-five years or so, has been MCPBA. The reluctance of suppliers to ship the *'rocket fuel'* high test hydrogen peroxide needs no explanation and regulations with respect to the transportation of pure MCPBA have also made the search for safe alternatives a sensible goal. MCPBA in its pure form is both shock-sensitive and potentially explosive in the condensed phase. The latter is a result of a positive result of 30 ms for MCPBA upon evaluation by the standard time/pressure test for products that are capable of deflagration [1]. The usual contaminant that is present in commercial MCPBA is *m*-chlorobenzoic acid and this may be removed by washing with a phosphate buffer of pH 7.5 and drying the residue under reduced pressure [2]. The presence of m-chlorobenzoic acid does lead to some reduction in the hazardous nature of MCPBA, but it is still shock-sensitive and capable of deflagration. By comparison, magnesium monoperoxyphthalate hexahydrate (MMPP) (1), a recently developed peroxygen product [3], is non-shock-sensitive and non-deflagrating. It is available both as a laboratory and as a bulk chemical [4]. Our early investigations showed that MMPP can be used in order to carry out a wide range of oxidation reactions, some of which we reported a few years ago [5]. We were also attracted to the possibility of using hydrogen bonded adducts of hydrogen peroxide by the report of the use of DABCO-di-perhydrate as an alternative to anhydrous hydrogen peroxide in the preparation of bis(trimethylsilyl) peroxide [6]. Our initial screening of perhydrates also included DABCO-di-*N*-oxide-di-perhydrate (2) [7], triphenylphosphine-oxide-perhydrate [8], and ureahydrogen peroxide (UHP) (3). The hydrogen bonded adduct (UHP), a white crystalline solid which is formed when urea is recrystallised from aqueous hydrogen peroxide [9], was investigated in detail for a number of reasons. It is very easily prepared and is also commercially available [4]. Infrared evidence favours the structure (3) in which hydrogen bonding occurs between a peroxide oxygen and one of the urea hydrogen atoms [10]. It is of interest to note that the complex that is formed between hydrogen peroxide and biuret is an inclusion complex that apparently does not involve strong hydrogen bonding of the type found in UHP.

(1)   (2)   (3)

Another feature that attracted us to UHP was the relatively high proportion of hydrogen peroxide (36.2%) in the adduct. The anhydrous material is hygroscopic and is best stored at low temperatures because of the possibility of thermal decomposition. However, the commercial material, which we use, typically has an available oxygen content of about 90% of the theoretical value. It may be stored at room temperature over long periods of time without significant loss of available oxygen. This material is evidently reasonably stable as judged by negative impact and pressure-time tests on small samples. A preliminary account of the use of UHP in a range of typical oxidation reactions has been published [11] and also the details of a new procedure for the preparation of bis(trimethylsilyl)peroxide [12].

Some of the early uses [13] of MCPBA indicate the initial objectives that one would consider when evaluating new potential substitutes. The epoxidation of alkenes such as 1-octene (Eq. 1) and ethyl crotonate (Eq. 2), and the epoxidation of enol ethers followed by cleavage (Eq. 3) [14] are illustrative.

$$\text{MCPBA in PhH} \atop 25\ ^{\circ}\text{C, 5 h, 81\%} \tag{1}$$

$$\text{MCPBA in DCE} \atop \Delta, 3\ \text{h, 70\%} \tag{2}$$

$$\text{MCPBA in CHCl}_3 \atop 25\ ^{\circ}\text{C, 17\%} \tag{3}$$

Baeyer-Villiger oxidation reactions, for example the oxidation of a 20-ketosteroid which afforded the 17-β-hydroxysteroid after hydrolysis of the ester [13], and the formation of N-oxides as shown in Eq. (4) [15] were also reported as early examples.

$$\text{MCPBA} \atop 21\% \tag{4}$$

2 Epoxidation Reactions

The successful examples of epoxidation reactions using MMPP have so far been carried out using a hydroxylic solvent to dissolve the reagent. This has usually been a low molecular weight alcohol or, where the substrate to be epoxidised is

3

insoluble in such a solvent, water has been used together with a phase tranfer catalyst to carry the monoperphthalate anion into an organic solvent such as chloroform. Thus 2-methylbut-2-ene (Eq. 5) and cyclohexene were both epoxidised in high yield using aqueous isopropanol as the solvent [5]. The epoxidation of 1,2-dimethylcyclohexa-1,4-diene (Eq. 6) with MMPP in ethanol [16] should be compared with the same reaction using MCPBA [17].

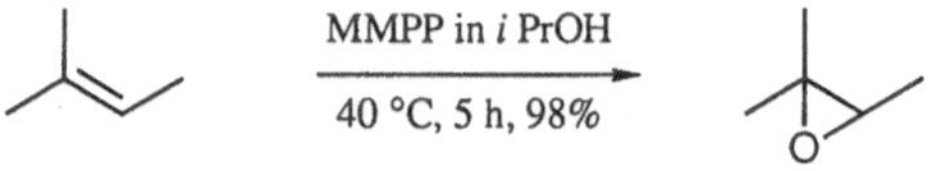

$$\text{MMPP in } i\text{ PrOH}, \quad 40\,°C, 5\,h, 98\% \tag{5}$$

$$\text{MMPP in EtOH}, \quad 42\% \tag{6}$$

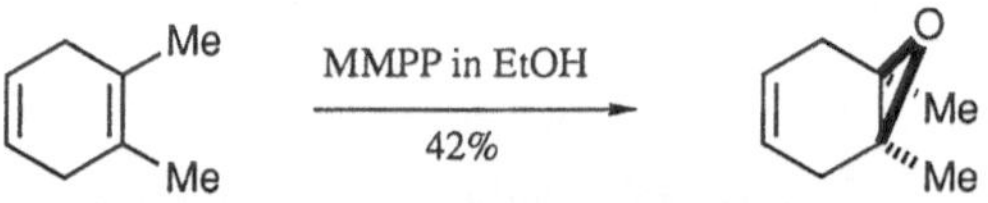

The epoxidation of dienes using MMPP occurs, as expected, at the more electron rich double bond, like those reactions using MCPBA. The example shown in Eq. (7) was one of the stages in a synthesis of the quinone antibiotic frenolicin and used MCPBA in the presence of 5% aqueous sodium hydrogen carbonate [18].

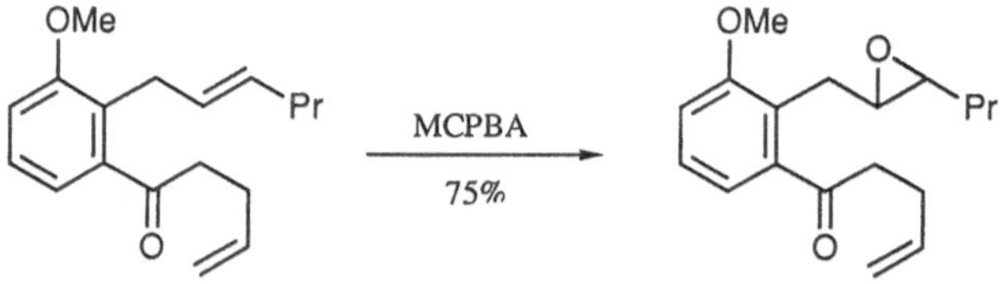

$$\text{MCPBA}, \quad 75\% \tag{7}$$

Thus limonene affords 1-methyl-4-(1-methylethenyl)-epoxycyclohexane (Scheme 1) using MMPP and the phase transfer reaction conditions. The procedures based

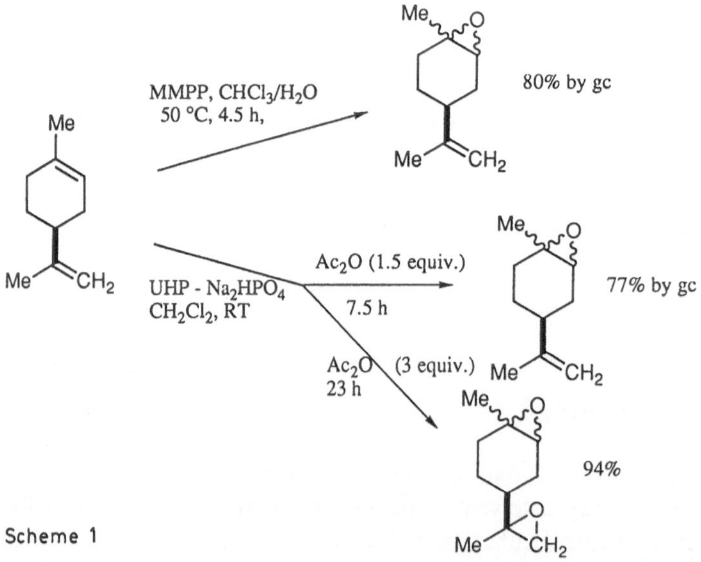

MMPP, CHCl$_3$/H$_2$O
50 °C, 4.5 h,
80% by gc

Ac$_2$O (1.5 equiv.)
UHP - Na$_2$HPO$_4$
CH$_2$Cl$_2$, RT
7.5 h
77% by gc

Ac$_2$O (3 equiv.)
23 h
94%

Scheme 1

on UHP that are presently available for carrying out epoxidation reactions on dienes using UHP do not allow good yields of monoepoxides to be isolated. This point is exemplified by the examples using UHP procedures shown in Scheme 1. It is worth noting, however, that as we would expect, geranyl acetate can be converted into 6,7-epoxygeranyl acetate. The epoxidation of enol ethers can be carried out successfully using a variety of reagents including MCPBA, peroxyimidic acids, and MMPP. A comparison of the reactions of ethoxyethene (Eq. 8) have been reported [19], as also has the sequence from cis-but-2-ene-1,4-diol to the corresponding epoxyester (Eq. 9) [20]. The reverse of the sequence shown in Eq. (9) gave low conversions.

$$\text{(8)}$$

$$\text{(9)}$$

There have been examples reported where attempted epoxidation reactions of alkenes have failed in the presence of MMPP. Terminal alkenes such as those shown in Eq. (10) did not afford an epoxide in attempted reactions using MMPP. However, as shown in Eq. (10) epoxidation was successful using MCPBA when using Kishi's high temperature method [21] in which a radical inhibitor, for example, 4,4'-thiobis-(6-*tert*-butyl-3-methylphenol), was employed [22]. The decomposition of peroxycarboxylic acids at high temperatures presumably occurs by radical processes. It may well be that some apparent failures using MMPP result from very slow rates of reactions that are carried out at room temperature. On the other hand, MMPP has been used successfully together with 5,10,15,20-tetra-2,6-dichlorophenylporphinatomanganese (III) acetate in oxygenation reactions that mimic cytochrome P-450 enzymes [23]. The addition of pyridine or 4-methylpyridine improves the rate of epoxidation reactions and even electron depleted alkenes such as isobutyl 3-butenoate are efficiently epoxidised under mild reaction conditions.

$$\text{(10)}$$

R = PhCH$_2$ or TMSCH$_2$CH$_2$

Selectivity in epoxidation reactions frequently depends on the reagent chosen and also on reaction conditions. The epoxidation of cholesterol using a range of peroxycarboxylic acids, including MMPP, gives a mixture in which the α-epoxide

5

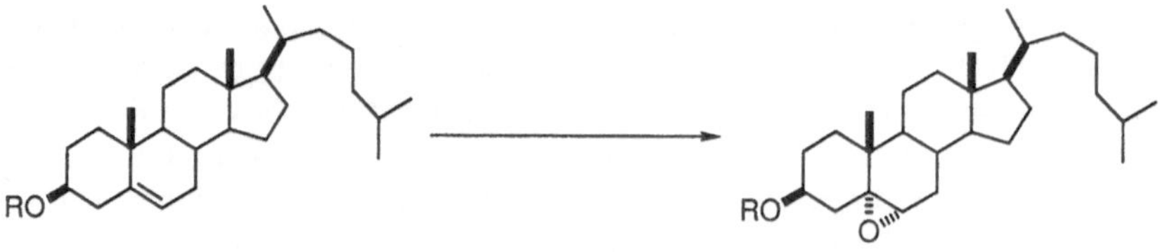

R = H	MMPP, CHCl$_3$, H$_2$O	R = H 88%	
R = H	UHP-(CF$_3$CO)$_2$O, Na$_2$HPO$_4$ at RT 16 h	R = CF$_3$CO 62%	ratio alpha : beta
R = PhCO	MMPP, CHCl$_3$, H$_2$O	R = PhCO	5.5 : 1
R = PhCO	UHP-(CF$_3$CO)$_2$O, Na$_2$HPO$_4$ at RT 16 h	R = PhCO 51%	3 : 1

Scheme 2

predominates over the β-epoxide by about 4:1 (Scheme 2). On the other hand, using trifluoroperoxyacetic acid generated by the UHP procedure [11] one obtains mainly the 3-β-trifluoroacetoxy-5,6-α-epoxide. The epoxidation of cholesteryl benzoate using MMPP gave a ratio of α- to β-epoxide of 5.5:1 and using trifluoroperoxyacetic acid an α- to β-epoxide ratio of ca. 3:1 was indicated by the ^{1}H NMR spectrum. It is of interest to note that the epoxidation of cholesteryl benzoate using the catalytic system shown in Eq. (11) gave an α:β ratio of ca. 3:7 [24]. This last result is reminiscent of the results obtained in reactions of steroidal alkenes with dioxiranes. Cholesteryl acetate reacts with dimethyldioxirane to afford a mixture of epoxides in which the β-isomer is present in larger amounts than is normally the case when using a peroxycarboxylic acid [25].

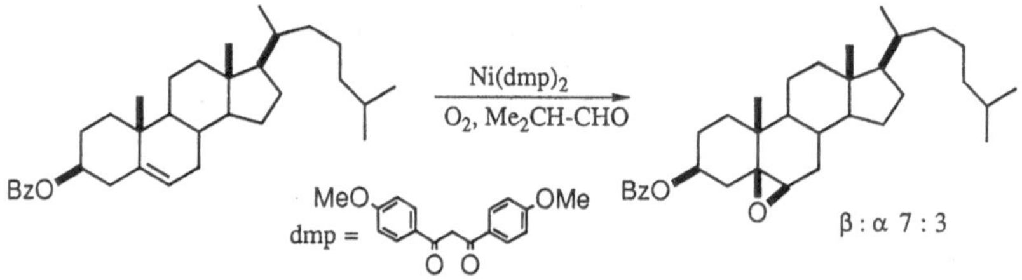

(11)

Selectivity in the epoxidation of dienes where the two alkene residues are a widely differing nucleophilicities is not difficult to achieve and we will return to this point later in this review. However, it is worth pointing out that quite good diastereofacial selectivity was reported in the epoxidation shown in Eq. (12), part of a total synthesis of (+)-altholactone. The epoxidation using MCPBA proceeded with poor selectivity whereas MMPP gave predominant attack on the β-face and a 3.5:1 mixture of the substituted tetrahydrofuran derivatives was obtained after acid catalysed cyclisation [26].

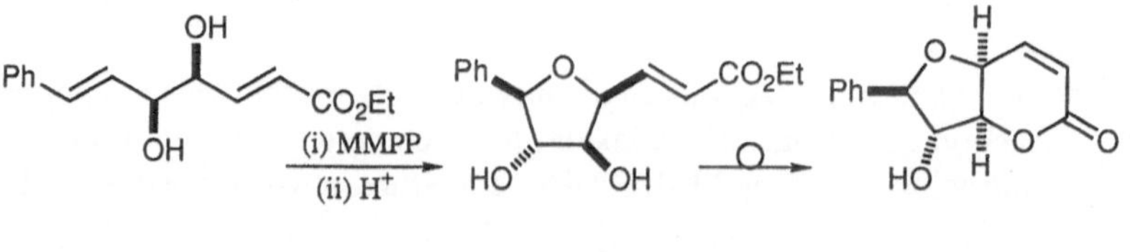

(12)

Epoxidation reactions carried using UHP as the source of hydrogen peroxide require that the anhydride and other reactions conditions are chosen carefully. For particularly nucleophilic alkenes our best procedure uses acetic anhydride and disodium hydrogen phosphate at room temperature. In this way we were able to obtain α-methylstyrene oxide in a 75% yield and the epoxide from α-pinene in 79% yield as shown in equation (13).

$$
\underset{\text{CH}_2\text{Cl}_2 \text{ RT 15 h, 79\%}}{\overset{\text{UHP - Na}_2\text{HPO}_4, \text{ Ac}_2\text{O,}}{\xrightarrow{\hspace{3cm}}}}
\tag{13}
$$

Magnesium monoperoxyphthalate has also been used in the epoxidation of vinylarenes, as shown in Eq. (14) [27].

$$
\overset{\text{MMPP, THF, H}_2\text{O}}{\xrightarrow{\hspace{2.5cm}}}
\tag{14}
$$

With the relatively non-nucleophilic terminal alkenes which are reasonably involatile, such as 1-octene, we have generated peroxytrifluoroacetic acid in the presence of disodium hydrogen phosphate and heated the mixture under reflux for a brief period. In this way we were able to obtain a high yield of 1,2-epoxyoctane as indicated in Eq. (15). Other examples of the use of these and other procedures are given in Table 1.

$$
\underset{(\text{CF}_3\text{CO})_2\text{O, } \Delta \text{ 0.5 h, 88\%}}{\overset{\text{UHP Na}_2\text{HPO}_4}{\xrightarrow{\hspace{3cm}}}}
\tag{15}
$$

The epoxidation of electron deficient alkenes such as methyl methacrylate has also been carried out using reaction conditions similar to those shown in Eq. (15), and with α,β-unsaturated ketones alkaline hydrogen peroxide has been generated from UHP and affords good yields of epoxides. Pulegone gave a 50% yield of the epoxide and the result obtained with isophorone is shown in Eq. (16).

$$
\underset{68\%}{\overset{\text{UHP-NaOH, MeOH}}{\xrightarrow{\hspace{2.5cm}}}}
\tag{16}
$$

The epoxidation of nitroalkenes such as β-methyl-β-nitrostyrene can be achieved using alkaline hydrogen peroxide [28]. The epoxide shown in Eq. (17) was obtained in almost quantitative yield when we used UHP in methanol-aqueous sodium

7

Table 1. Epoxidation reactions using urea-hydrogen peroxide (UHP)

Substrate	Product	Method	Yield [%]
styrene	styrene oxide	(1)	60
α-methylstyrene	1-phenyl-1-methyloxirane	(1)	83
β-methylstyrene	1-phenyl-2-methyloxirane	(1)	86
trans-stilbene	1,2-diphenylepoxyethane	(1)	47
2,3-dimethylbut-2-ene	2,3-dimethyl-2,3-epoxybutane	(1)	51
hex-1-ene	1,2-epoxyhexane	(2)	53
oct-1-ene	1,2-epoxyoctane	(3)	88
cyclohexene	epoxycyclohexene	(1)	74
1-methylcyclohexene	1-methylepoxycyclohexane	(1)	56
3-methylcyclohexene	3-methylepoxycyclohexane	(1)	58
limonene	limonene diepoxide	(1)	94
α-pinene	1,2–α-epoxypinane	(1)	79
α-ionone	1,1,3-trimethyl-2-(3-oxo-but-1-ethyl)-3,4-epoxycyclohexane	(1)	90
cyclo-octene	epoxycyclo-octane	(3)	60
phenyl allyl ether	1,2-epoxy-3-phenoxypropane	(2)	51
linalool	2-methyl-5-(2′-propyl-2′-hydroxy)-2-vinyl-tetrahydrofuran †	(2)	64
geraniol	geraniol-diepoxide	(1)	66
geranyl acetate	6,7-epoxygeranyl acetate	(1)	82
cholesterol	3-β-5,6-α-epoxycholestane and 3-β-5,6-β-epoxycholestane	(1)	65 / 28
cholesterol	3-β-trifluoroacetoxy-5,6-α-epoxycholestane	(2)	62
cholesteryl benzoate	3-β-benzoyloxy-5,6-α-epoxycholestane^ʃ	(2)	53
isophorone	3,5,5-trimethyl-2,3-epoxycyclohexanone	(4)	68
pulegone	pulegonepoxide	(4)	50
α-ionone	1,1,3-trimethyl-2-(3-oxo-1,2-epoxybutyl)-cyclohex-3-ene	(4)	68
β-methyl-β-nitrostyrene	2-nitro-3-phenyl-2,3-epoxypropane	(4)	94
methyl methacrylate	methyl 2-methyl-2,3-epoxypropanoate	(3)	56

† major product; ^ʃ ratio α : β 5 : 2;
methods (1) UHP-Ac$_2$O-Na$_2$HPO$_4$, 0 °C-RT; (2) UHP-(CF$_3$CO)$_2$O-Na$_2$HPO$_4$; RT; (3) UHP-(CF$_3$CO)$_2$O-Na$_2$HPO$_4$, reflux; (4) UHP-NaOH/MeOH.

hydroxide at low temperatures. At higher temperatures and in the presence of stronger than 2 mol l^{-1} sodium hydroxide we obtained mixtures of benzyldehyde and benzoic acid which result from hydroxide ion induced fragmentation of the epoxide and further oxidation.

Ph—CH=C(NO$_2$)(Me) $\xrightarrow[\text{0-5 °C, 94\%}]{\text{UHP-2mol l}^{-1}\text{ NaOH, MeOH}}$ Ph—epoxide—NO$_2$(Me) (17)

As expected the epoxidation of α-ionone can be carried out using different procedures in order to functionalise the two different double bonds as shown in Scheme 3.

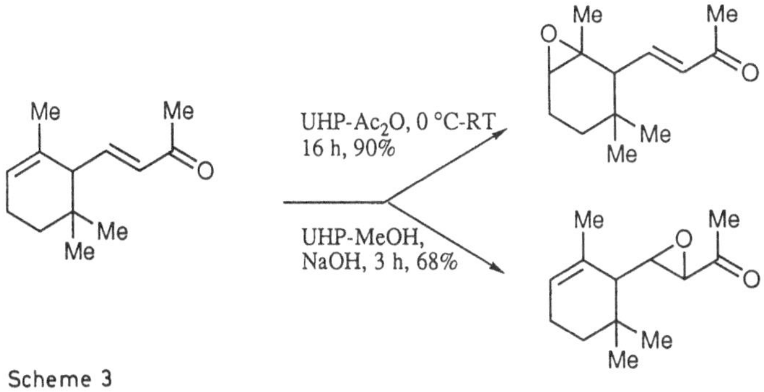

Scheme 3

We mentioned earlier the epoxidation of geranyl acetate using the UHP method. Conventional epoxidation of geraniol using MCPBA results in the formation of a 2:1 mixture of the 6,7-epoxide and the 2,3-isomer despite the reduction in electron density at the 2,3-position caused by the allylic hydroxyl group [29]. In this connection it is interesting to note that the 2,3-epoxide is formed in a 93% yield by using an emulsion technique in which the 6,7-double bond is kept away from the MCPBA in a hydrocarbon phase [30].

The oxidation of some heteroaromatic compounds have been reported using MMPP. Some of the products clearly result from initial epoxidation. The reaction shown in Eq. (18) was used in the synthesis of bromobeckerelide, a product isolated from a marine red alga [31]. Other furan derivatives that carry electron releasing groups in the 2- and 5-positions afford ring opened dienones using either MCPBA or MMPP. In the latter case the reactions were carried out in aqueous ethanol and gave very good yields [32]. The epoxidation of 1-benzenesulfonyl-indole which results in the formation of the corresponding indoxal is shown in Eq. (19) [33].

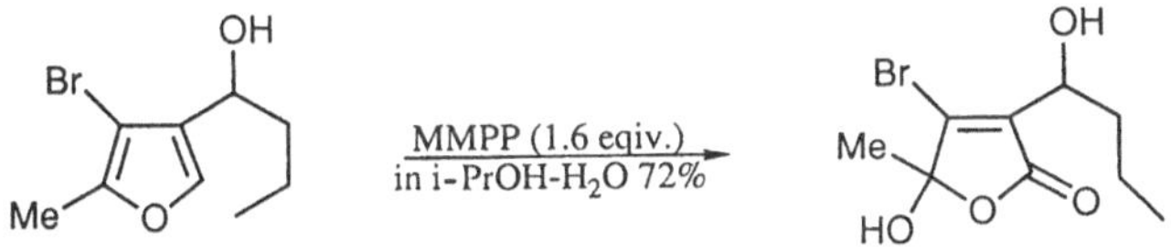

$$\text{(18)}$$

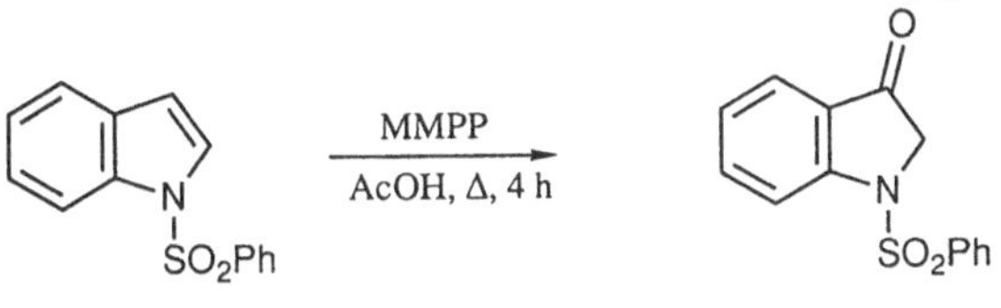

$$\text{(19)}$$

Before leaving the subject of epoxidation reactions we should make brief mention of the control of stereochemistry involving the use of neighbouring functional

$$\text{(20)}$$

groups. The reaction shown in Eq. (20) involves the use of MCPBA [34]. High diastereofacial selectivity has also been observed in epoxidation reactions using MMPP. Both of the diastereomers of the phosphine oxides shown in Eq. (21) are epoxidised with stereoselectivities in excess of 95:5 [35]. Reactions using compounds which just retained the chiral phosphorus centre gave very low selectivities and show that the stereoselectivity is dominated by the effect of the chiral carbon centre. It was assumed that the stereochemical control resulted from Houk selectivity [36] as indicated in structure (4). Similar arguments have been adduced in connection in a number of reactions of allyl silanes, including epoxidation reactions [37]. The osmium tetroxide catalysed hydroxylation of E-γ-hydroxy-α,β-unsaturated esters using N-methylmorpholine-N-oxide and aqueous acetone has also been explained in terms of a conformation resulting from favourable interactions between the p-orbitals of the double bond and unshared electron pair on the γ-oxygen [38].

$$\text{(21)}$$

An = 2-MeO-C$_6$H$_4$

3 Bayer-Villiger Oxidations

The facility with which the oxidation of ketones can be effected is related to the strength of the conjugate acid of the leaving group. The stronger the acid the more powerful is the peroxyacid in its reactions. Peracids that are commonly available, for example MCPBA, are the most widely reported. Trifluoroperoxyacetic acid is a remarkably powerful reagent for Bayer-Villiger reactions [39] but the former requirement for the availability high strength hydrogen peroxide has limited its use. On the other hand there have been examples reported where the use of lower strength hydrogen peroxide has been successful [40]. Two examples illustrate recent cases where high strength trifluoroperoxyacetic acid, prepared using 90% hydrogen peroxide, and MCPBA have been used. Equation (22) shows how a Baeyer-Villiger reaction has been used as part of stereocontrolled general synthesis of C-nucleosides [41], and the use of MCPBA in the conversion of adamantanone into 2-thiaadamantane is shown in Eq. (23) [42].

In our preliminary report on the use of MMPP we mentioned two examples where Baeyer-Villiger reactions had been carried out satisfactorily [5]. Pinacolone was converted into-tert-butyl acetate and cyclohexanone into caprolactone in good yields. There have been a small number of other reports of the use of MMPP. The reactions of the β-lactams shown in Eq. (24) are reported to proceed efficiently using MMPP [43]. However, although the cubane shown in Eq. (25) undergoes the tris-Bayer-Villiger reaction in good yield using MCPBA, an attempted reaction using MMPP failed. On the other hand, the use of the UHP-trifluoroacetic anhydride method was found to give an 80% yield of the expected product [44].

R^1 = H, TMS, or TBDMS
R^2 = H, PhCH$_2$, or 4-MeOC$_6$H$_4$CH$_2$
R^3 = Ph or Me

A = C(OH)MeBu^t A = C(OH)MeBu^t

One of the first examples of Baeyer-Villiger reactions to be reported involved the oxidation of menthone using Caro's acid. The Baeyer-Villiger oxidation of cycloalkanones (Eq. 26) using trifluoroperoxyacetic acid generated by the interaction of trifluoroacetic anhydride with UHP using disodium hydrogen phosphate as a buffer results in the formation of the expected lactones in good yields. The exclusive formation of 7-methylcaprolactone from 2-methylcyclohexanone illustrates the generalisation that secondary alkyl groups have higher migratory aptitudes than a primary alkyl group, as also does the isolation in very high yield of a single product from menthone.

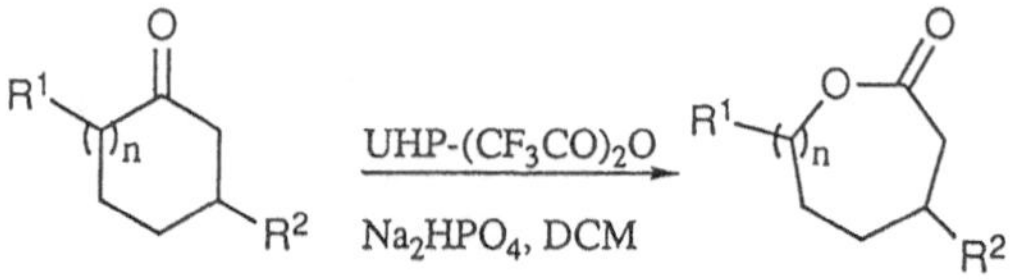

$$
\begin{array}{ll}
n = 0, R^1 = H & 53\% \\
n = 1, R^1 = H & 61\% \\
n = 1, R^1 = Me & 63\% \\
n = 1, R^1 = Pr^i, R^2 = Me & 98\% \\
n = 2, R = H & 85\%
\end{array}
$$

(26)

The method also works well for bicyclic ketones such as nor-camphor (Eq. 27) and acyclic ketones such as pinacolone which afforded *tert*-butyl acetate in 96% yield. Reactions of benzocycloalkanones such as α-tetralone with the UHP-trifluoroacetic anhydride system also give the expected lactones in better yields than have previously been reported. The results are summarised in Eqs. (28) to (30).

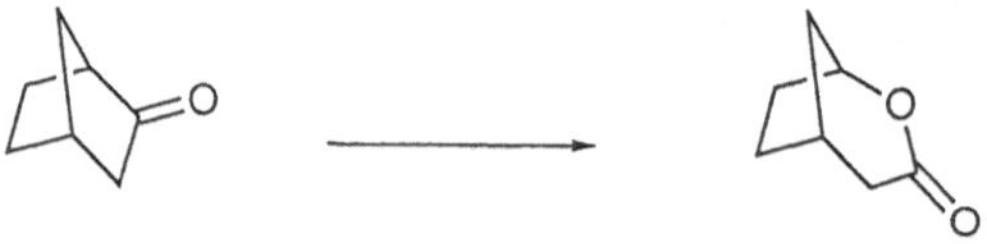

(27)

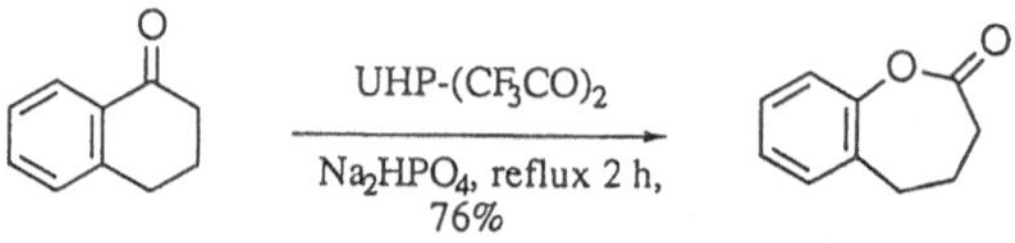

UHP-(CF$_3$CO)$_2$

Na$_2$HPO$_4$, reflux 2 h, 76%

(28)

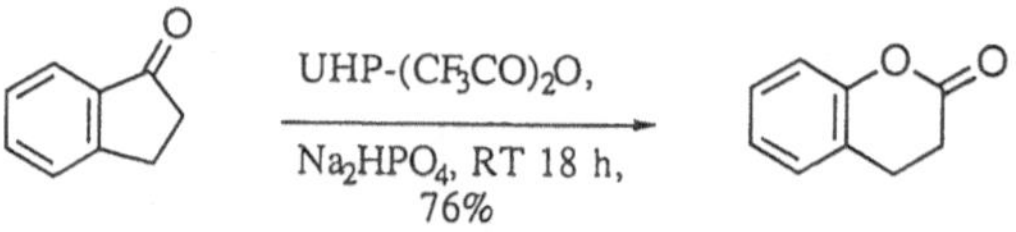

UHP-(CF$_3$CO)$_2$O,

Na$_2$HPO$_4$, RT 18 h, 76%

(29)

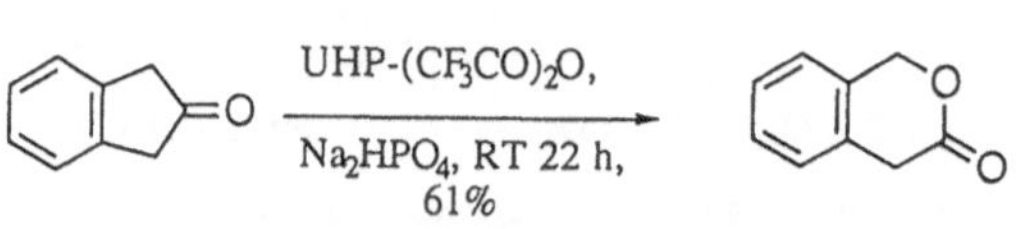

UHP-(CF$_3$CO)$_2$O,

Na$_2$HPO$_4$, RT 22 h, 61%

(30)

As expected, Baeyer-Villiger reactions of simple arylalkyl ketones are oxidised to esters in which aryl migration is observed. Examples are shown in Eq. (31) to (33).

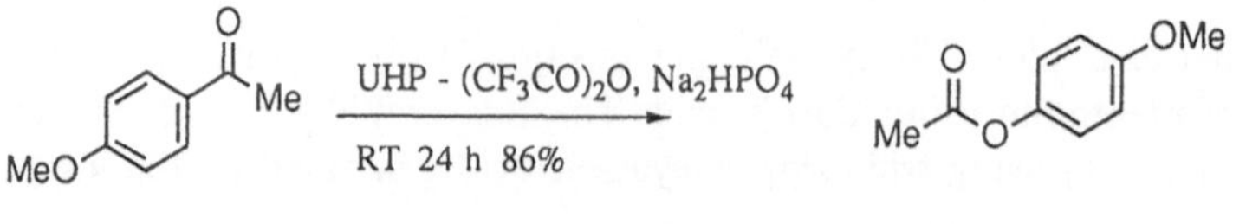

UHP - (CF$_3$CO)$_2$O, Na$_2$HPO$_4$

RT 24 h 86%

(31)

12

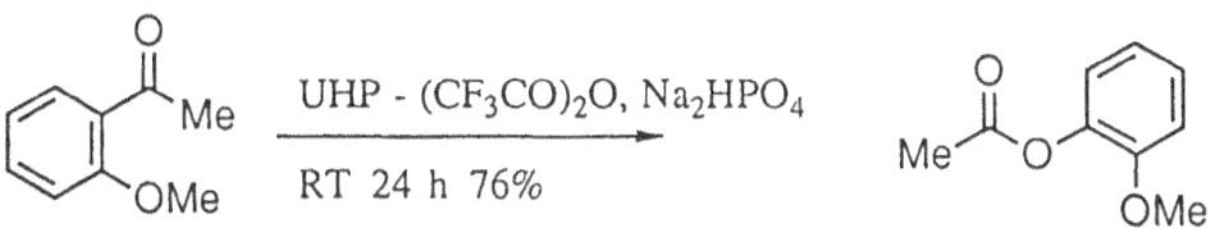

$$\text{(32)}$$

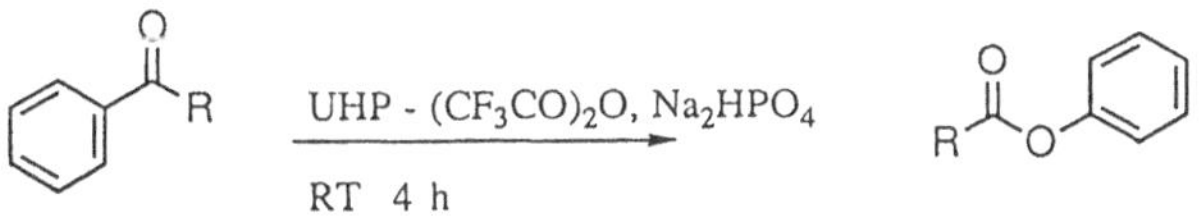

$$\text{(33)}$$

R = Me,   53%
R = Et,   94%
R = Pr^n,   86%
R = Bu^n,   86%

The oxidation of aromatic aldehydes to phenols via the aryl formates is known as the Dakin reaction and is evidently related to the Baeyer-Villiger oxidation of ketones. The use of MCPBA is well known [45]. The successful use of MMPP and UHP-acetic anhydride in these reactions (Eqs. 34 to 36) can be achieved with aromatic aldehydes that have an electron releasing substituent in an *ortho-* or *para*-position. In the absence of a suitable electron releasing substituent, aromatic aldehydes are oxidised to the corresponding carboxylic acid.

In fact the use of UHP and methanolic sodium hydroxide allows the oxidation of aromatic aldehydes to the corresponding acids to be carried out in good yields. Some examples are shown in Eq. (37).

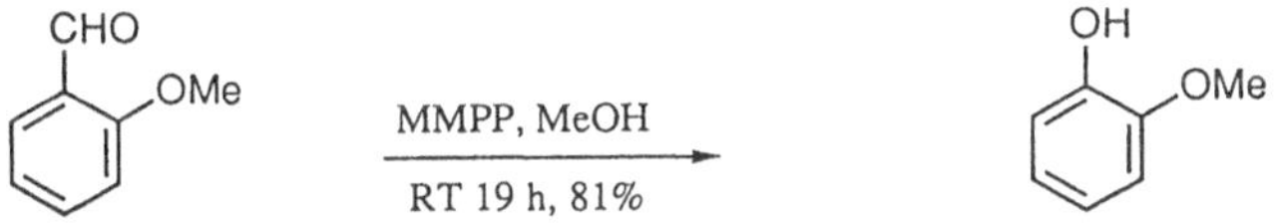

$$\text{(34)}$$

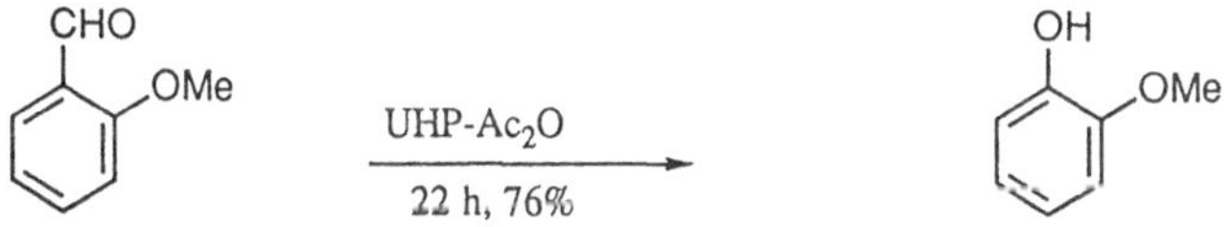

$$\text{(35)}$$

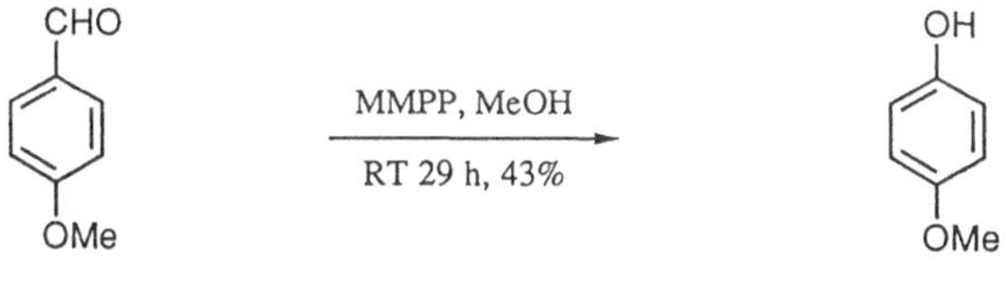

$$\text{(36)}$$

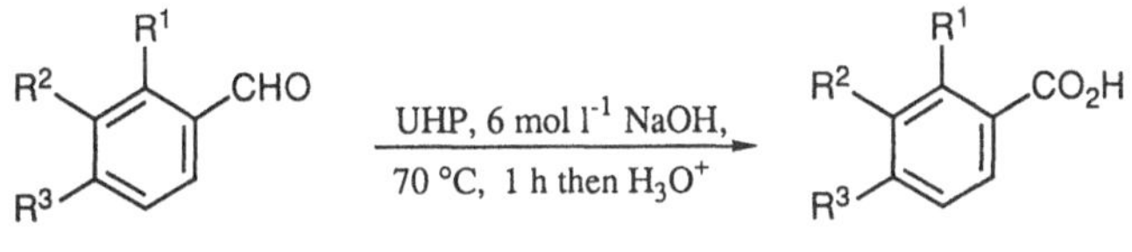

$$\text{(37)}$$

$R^1 = R^2 = R^3 = H$ 94%
$R^1 = MeO, R^2 = R^3 = H$ 90%
$R^1 = R^3 = H, R^2 = MeO$ 95%
$R^1 = R^2 = H, R^3 = MeO$ 72%
$R^1 = H, R^2 = R^3 = MeO$ 63%

4 Heteroatom Oxidations

4.1 The Oxidation of Sulfur

The oxidation of sulfides to sulfoxides or sulfones depend on the reaction conditions used. Hydrogen peroxide and *tert*-butyl hydroperoxide have both been used as also has a variety of peroxycarboxylic acids. Since all of the peroxycarboxylic acids are stronger oxidants than hydrogen peroxide, reagents such as MCPBA have been used to oxidise sulfides to sulfoxides under very mild conditions. In the example shown in Eq. (38) the diastereomeric sulfoxides were obtained in good yield [46].

$$\text{(38)}$$

In our preliminary report on the use of **MMPP** we showed that tetra-hydrothiophene could be oxidised selectively to afford either the sulfoxide or the sulfone [5], and using UHP at room temperature we were only able to isolate the sulfone [11]. Careful control of the amount of **MMPP** used has allowed the preparation of sulfoxides as shown in Eqs. (39 and 40) [47], while use of an excess of **MMPP** allows sulfones (Eqs. 41 and 42) to be isolated in good yields [48, 49].

$$\text{(39)}$$

$$\text{(40)}$$

$$\text{(41)}$$

$$\text{(42)}$$

4.2 The Oxidation of Nitrogen

Pyridine, 2-methylpyridine, and 2-chloropyridine can all be oxidised to the corresponding N-oxides using MMPP in acetic acid [5]. In our more recent work we have also shown that quinoxaline can be oxidised to the di-N-oxide using the UHP-trifluoroacetic anhydride procedure in moderate yield [11]. It is likely that isolation difficulties were a major problem in the latter case. The oxidation of quinoxaline using MMPP to the di-N-oxide in 1,2-dichloroethane proceeds in high yield (Eq. 43) and there are then no difficulties in isolating the product. The oxidation of the pyrimidine shown in Eq. (44) has been reported [50] using MMPP and a yield of 91% obtained in the oxidation shown in Eq. (45) [51].

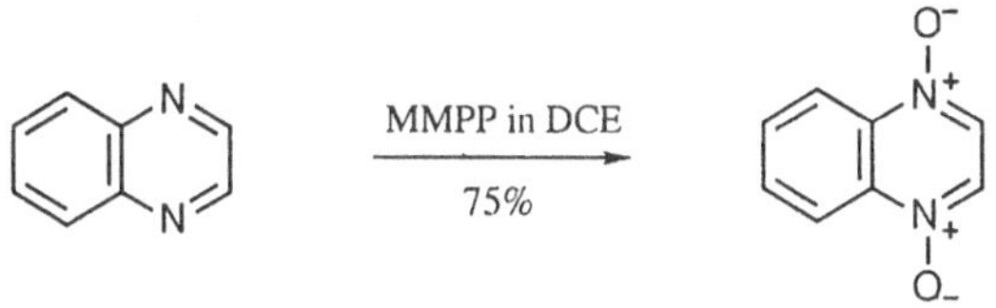

(43)

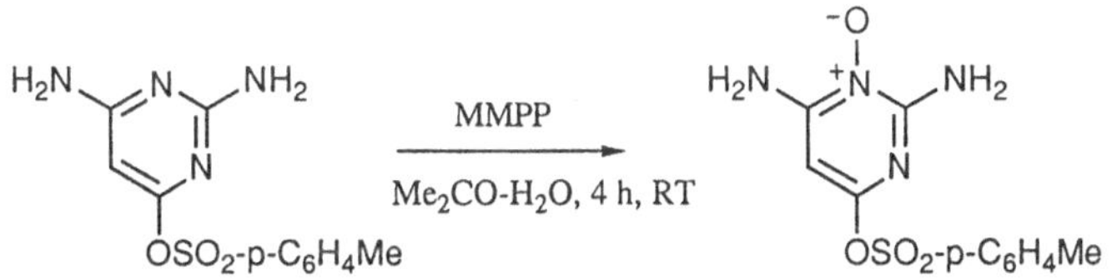

(44)

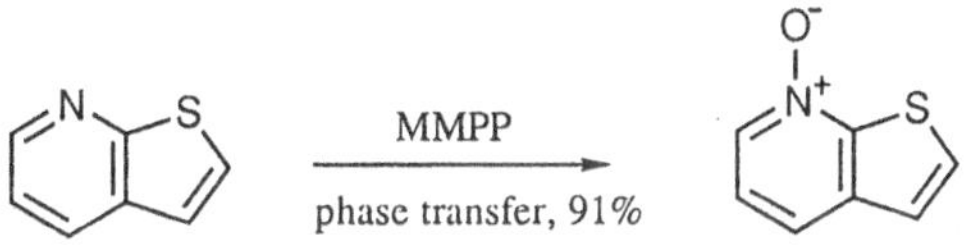

(45)

When we started our study of the use of MMPP and UHP we were unaware of an interesting report of the use of UHP as a substitute for 90% hydrogen peroxide in N-oxidations using the system that consisted of a mixture of trifluoroacetic acid, concentrated sulphuric acid, and hydrogen peroxide [52]. The oxidation of the 1,8-naphthyridine derivative shown in Eq. (46) indicates the merit of that method.

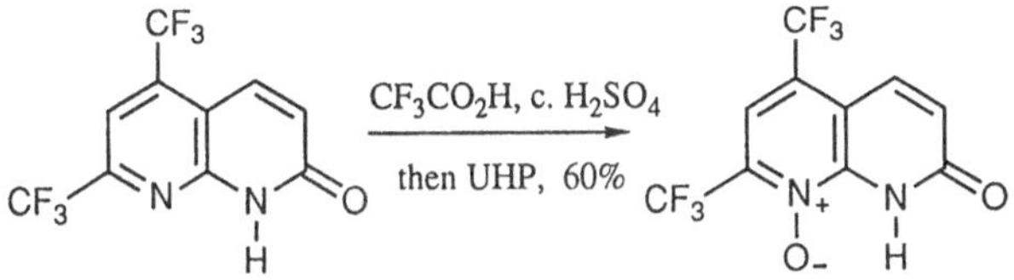

(46)

Dipolar cycloaddition reactions of nitrones, prepared by the oxidation of suitable hydroxylamines, have been used in a number of alkaloid syntheses. The preparation of tetrahydropyridine N-oxides by means of the oxidative cleavage of suitable bicyclic isoxazolidines has been studied in connection with projected syntheses of indolizidine alkaloids [53]. The oxidation of the isoxazolidine (5) afforded a mixture of the nitrones (6) and (7) in which the compound (7) was slightly favoured in reactions using MCPBA [ratio of (6):(7) ca. 1:1.4]. On the other hand, MMPP gave more useful amounts of the more substituted nitrone (7) particularly when the reaction is carried out at low temperatures [Eq. (47)]. Oxidation of the isoxazolidine (8) with MCPBA [Eq. (48)] gave the nitrones (9) and (10) in which the product (9) predominated. In contrast, the regiochemical control was superior using MMPP and gave the nitrone (10) in 68% yield, as the only product, when the oxidation was carried out in methanol at 20 °C.

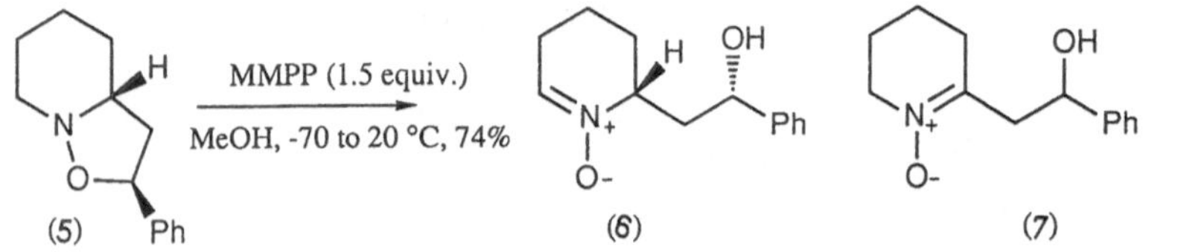

$$\text{ratio } (6):(7) = 5.9:1 \tag{47}$$

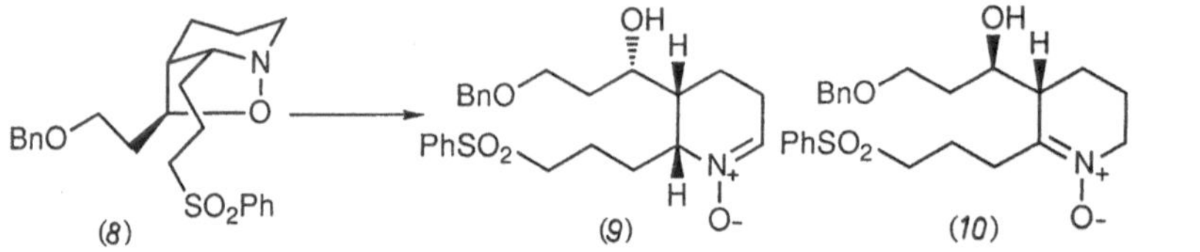

(48)

One of the potentially most valuable uses for MMPP that have been developed is connected with the oxidative regeneration of ketones from hydrazones. Dimethyl-hydrazones may be easily metallated and these derivatives are thus valuable enolate equivalents in carbon-carbon bond forming reactions. This methodology has been extended by means of the SAMP-/RAMP-hydrazone procedures which thereby allows high regio-, diastereo-, and enantioselective electrophilic substitution reactions to be effected α- to the original carbonyl group [54]. A number of methods are available for the cleavage of dimethylhydrazones including the use of sodium perborate [55]. Very high yields have been reported for the oxidative cleavage of dimethyl- and SAMP-hydrazones using MMPP, in the latter cases essentially no racemisation was observed with a chiral centre α- to the carbonyl group [56]. A generalised scheme is shown in Eq. (49) and two specific examples in Eq. (50) and (51).

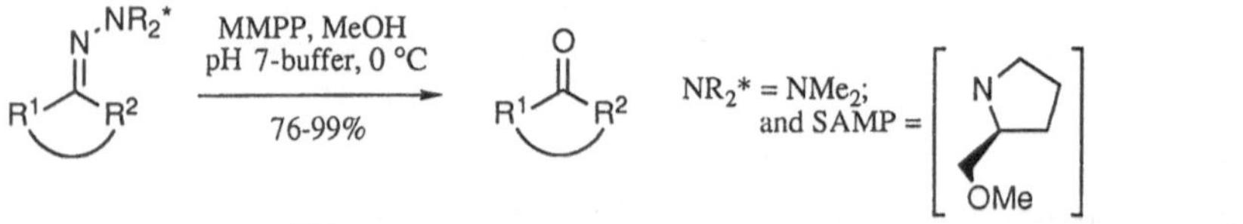

(49)

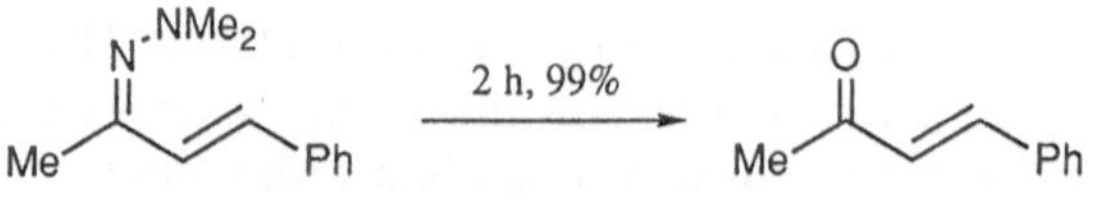

(50)

$$\text{C}_{11}\text{H}_{23} \overset{\text{S}}{\underset{\text{Me}}{\bigg|}} \overset{\text{N}^{\diagup}\text{N}}{\underset{\text{Ph}}{\bigg\|}} \text{OMe} \quad \xrightarrow[\text{> 96\% ee}]{\text{2 h, 82\%}} \quad \text{C}_{11}\text{H}_{23} \overset{\text{S}}{\underset{\text{Me}}{\bigg|}} \overset{\text{O}}{\underset{}{\bigg\|}} \text{Ph} \qquad (51)$$

In our work using MMPP and UHP systems we have compared the efficiency of the oxidative cleavage of dimethyl- and phenylhydrazones. In the case of dimethylhydrazones we have observed that the MMPP method is normally better. For example, although the dimethylhydrazone derived from acetophenone gave compararable results using MMPP and UHP-Trifluoroacetic anhydride-disodium hydrogen phosphate, 83% and 77% yields of isolated products respectively, the results obtained with the derivatives of aliphatic ketones were markedly better using MMPP. The cleavage reactions of phenylhydrazones were effected in slightly better yields using the UHP method as indicated in Eq. (52).

$$\underset{\text{Me}}{\overset{\text{N}^{\diagup\text{NHPh}}}{\bigg\|}}\text{Ph} \quad \xrightarrow[\text{UHP, 67\%}]{\text{MMPP, 37\%}} \quad \underset{\text{Me}}{\overset{\text{O}}{\bigg\|}}\text{Ph} \qquad (52)$$

5 Conclusions

I hope that I have shown in this review that it has been possible to develop new reagent systems that go some way to overcoming the perceived disadvantages of established methodology. The reagents that I have discussed have been used with complete safety which is a very important consideration. That is not to say that we should ever take risks with peroxidic reagents — care must continue to be taken when carrying out reactions with magnesium monoperoxyphthalate and urea-hydrogen peroxide. It is also clear that a number of colleagues are also taking advantage of the reagents to develop improved procedures and improved selectivities. The number of citations, especially to the use of magnesium monoper-oxyphthalate, is increasing each year. Recent examples of the use of MMPP [57, 58, 59] include the oxidation of sulphide to sulphoxide as part of a stereoselective synthesis of the antimalarial compound (+)-artimisinin [59]. In addition the range of oxidation reactions that can be effected has also widened. Other colleagues are also begining to find some advantage in the use of UHP [60, 61, 62]. A UHP method has been used as a reagent in the synthesis of the brassinolide side chain [61], and in a GoAgg^{II} reaction UHP has been used as an alternative to anhydrous hydrogen peroxide [62]. There are still problems to be resolved but the way forward is clear and it is anticipated that reagents of the type reviewed will be developed within the next few years that will progress the control of oxidation processes.

Acknowledgements. It is a pleasure to be able to thank my friends at Interox (Widnes) for their support, especially Paul Brougham, David Cummerson, Bill Sanderson and Phil. Sankey. Finally, it is important to note that without the help and valuable experimental skills of my co-workers mentioned in the references it would not be possible for me to write this lecture-review. I am grateful to them for all of their efforts. Where unpublished results are reported here the work has been carried out by Amanda Newbold as part of her Ph D programme supported by SERC and Interox (Widnes) through the CASE scheme.

6 References

1. UN test method, Test 2 (b) (iii) – ST/AG/AC 10/11
2. Schwartz NN, Blumbergs JH (1964) J Org Chem, 29: 1976
3. European Patent Appl 27693 (1981); Chem Abs (1981) 95: 168801
4. Available from Interox Chemicals Ltd, Baronet Works, P.O. Box 7, Warrington, Cheshire, WA46HB, U.K. Also from Aldrich Chemical Company Ltd
5. Brougham P, Cooper MS, Cummerson DA, Heaney H, Thompson N (1987) Synthesis 1015
6. Cookson PG, Davies AG, Fazal N (1975) J Organometal Chem, 99: C31
7. Oswald AA, Guertin DL (1963) J Org Chem, 28: 651–657
8. Temple RD, Tsuno Y, Leffler JE (1963) J Org Chem 28: 2495; Copley DB, Fairbrother F, Miller JR, Thompson A (1964) Proc Chem Soc, 300–301
9. Lu C-S, Hughes EW, Giguère PA (1941) J Am Chem Soc, 63: 1507–1513
10. Aida K (1963) J Inorg Nucl Chem, 25: 165–170
11. Cooper MS, Heaney H, Newbold AJ, Sanderson WR (1990) 533–535
12. Jackson WP (1990) Synlett 536
13. Fieser LF, Fieser M (1967) Reagents for organic synthesis, vol 1, J Wiley, New York, pp 135–139
14. Borowitz IJ, Gonis G (1964) Tetrahedron Lett, 1151–1155
15. Delia TJ, Olsen MJ, Brown GB (1965) J Org Chem, 30: 2766–2768
16. Gillard JR, Newlands MJ, Bridson JN, Burnell DJ (1991) Canad J Chem, 69: 1337–1343
17. Paquette LA, Barrett JH (1973) Org Synth, Coll Vol V, 467–471
18. Semmelhack MF, Zask A (1983) J Am Chem Soc, 105: 2034–2043
19. Machida S, Hashimoto Y, Saigo K, Inoue J-y, Hasegawa (1991) 47: 3737–3752
20. Grandjean D, Pale P, Chuche (1991) Tetrahedron Lett, 32: 3043–3046
21. Kishi Y, Aratani M, Tanino H, Fukuyama T, Goto T, Inoue S, Sugiura S, Kakoi H (1972) J Chem Soc, Chem Commun, 64–65
22. Petter RC (1989) Tetrahedron Lett, 30: 399–402
23. Querci C, Ricci M (1989) J Chem Soc, Chem Commun, 889–890
24. Yamada T, Takai T, Rhode O, Mukaiyama T (1991) Bull Chem Soc Jpn, 64: 2109–2117
25. Marples BA, Muxworthy JP, Baggaley KH, (1991) Tetrahedron Lett, 32: 533–536
26. Kang SH, Kim WJ (1989) Tetrahedron Lett, 30: 5915–5918
27. Chang HM, Chui KY, Tan FWL, Yang Y, Zhong ZP (1991) J Med Chem, 34: 1675–1692
28. Newman H, Angier RB (1970) Tetrahedron, 26: 825–836
29. Sharpless KB, Michaelson RC, (1973) J Am Chem Soc, 95: 6136–6137
30. Nakamura M, Tsutsuki N, Takeda T, Rokroyama T, (1984) Tetrahedron Lett, 25: 3231–3232
31. Jefford CW, Jaggi D, Boukouvalas J (1989) Tetrahedron Lett, 30: 1237–1240
32. Dominguez C, Csáky AG, Plumet J (1990) Tetrahedron Lett, 31: 7669–7670.
33. Conway SC, Gribble GW (1990), Heterocycles 30: 627–633
34. Kocovsky P (1988) Tetrahedron Lett, 29: 2475–2478
35. Harmat NJS, Warren S (1990) Tetrahedron Lett, 31: 2743–2746
36. Houk KN, Moses SR, Wu Y-D, Rondan NG, Jäger V, Schohe R, Fronczek FR (1984) J Am Chem. Soc, 106: 3880–3882
37. Fleming I, Sarkar AK, Thomas AP (1987) J Chem Soc, Chem Commun, 157–159
38. Stork G, Kahn M (1983) Tetrahedron Lett, 24: 3951–3954

39. Lewis SN (1969) In: Augustine RL Oxidation (ed) vol 1, Dekker, New York, pp 213–258
40. Anastasia M, Allevi P, Ciuffreda P, Fiecchi A, Scala A (1985) J Org Chem, 50: 321–325
41. Noyori R, Sato T, Hayakawa Y (1978) J Am Chem Soc, 100: 2561–2563
42. Suginome J, Yamada S (1986) Synthesis, 741–743
43. Ricci M, Altamura M, Bianchi D, Cobri W, Gotti N (1988) to Farmitalia Carlo Erba, Spa, GB Pat 2196340
44. Professor Eaton PE, (February 1991) University of Chicago, Personal Communication
45. Godfrey IM, Sargent MV, Elix JA (1974) J Chem Soc, Perkin Trans 1, 1353–1354
46. van den Broek LAGM, in't Veld PJA, Colstee JH, Ottenheijm HCJ (1989) Synth Commun, 19: 3397–3405
47. Batty D, Crich D, Fortt SM (1990) J Chem Soc, Perkin Trans. 1 2875–2879
48. Crich D, Ritchie TJ (1990) J Chem Soc, Perkin Trans. 1 945–954
49. Siemens LM, Rottnek FW, Trzupek LS (1990) J Org Chem, 55, 3507–3511
50. Lamsa J, to Oy Farmos-Yhtymä (1987) Euro Pat 0270201
51. Klemm LH, Wang J, Sur SK (1990) J Heterocycl Chem, 27: 1537–1541
52. Eichler E, Rooney CS, Williams HWR (1976) J Heterocycl Chem, 13: 41–42
53. Holmes AB, Hughes AB, Smith AL (1991) Synlett 47–48
54. Enders D (1984) In: Morrison JD (ed) Asymmetric synthesis, vol 3 Academic, Orlando, pp 275–339
55. McKilop A, Tarbin JA (1987) Tetrahedron, 43: 1753–1758
56. Enders D, Plant A (1990) Synlett 725–726
57. Wagner J, Vogel P (1991) Tetrahedron, 47: 9641–9658
58. Forbes JE, Bowden MC, Pattenden G (1991) J chem Soc, Perkin Trans, 1: 1967–1973
59. Avery MA, Chong WKM, Jennings-White C (1992) J Am Chem Soc, 114: 949–979
60. Gonsalves AMR, Johnstone RAW, Pereira MM, Shaw J (1991) J Chem Res S, 208–209
61. Back TG, Blazecka PG, Krishna MV (1991) Tetrahedron Lett, 4817–4818
62. Barton DHR, Bévière SD, Chavasiri W, Csuhai E, Doller D, Liu W-G (1992) J Am Chem Soc, 114: 2147–2156

Homogeneous and Heterogeneous Catalytic Oxidations with Peroxide Reagents

R. A. Sheldon

Faculty of Chemical Technology and Materials Science, Laboratory of Organic Chemistry, P.O. Box 5045, 2600 GA Delft, The Netherlands

Table of Contents

Topics in Current Chemistry, Vol. 164
© Springer-Verlag Berlin Heidelberg 1993

Increasingly stringent environmental constraints are providing the stimulus for the replacement of traditional stoichiometric oxidants, e.g. dichromate, permanganate, with cleaner catalytic technologies in industrial organic synthesis. Catalytic oxygen transfer employing peroxide reagents as the oxygen donor provides an attractive alternative. Virtually all of the transition metals and several main group elements can be used in combination with peroxides, e.g. H_2O_2, RO_2H, to effect a plethora of organic transformations.

A further consideration involves the choice of a homogeneous or heterogeneous catalyst. The ideal catalyst is one that combines the high activity and selectivity associated with the former with the stability and ease of recovery of the latter.

The scope and potential of recent developments in this area are reviewed, with particular emphasis on the site-isolation of redox metals in the lattices of molecular sieves and enantioselective oxidations employing chemo- or biocatalysts.

1 Introduction to Clean Technologies

The last decade of the twentieth century seems destined to go down in history as the age of "environmentality". This is evident both in the general trends in society as a whole and in the chemical industry in particular. Some of these trends are summarized in Fig. 1. A direct consequence of these trends is a higher degree of technological specialization amongst chemical manufactures. As forward-looking manufacturers aspire to a higher level of technological sophistication it is better to excel in a few key technologies than to be mediocre in a wide range.

- Cleaner, more environmentally acceptable products, e.g. "green gasoline" (lead- and aromatics-free)
- Products that are more effective, more targeted in their action and environmentally friendly, i.e. readily recycled or biodegraded
- Zero emission plants/integrated waste management
- Cleaner technologies with negligible inorganic salt formation (i.e. catalytic processes with optimal *atom utilization*)
- Replacement of toxic and/or hazardous reagents, e.g. $COCl_2$, $(CH_3)_2SO_4$, H_2CO/HCl, heavy metal salts. Transport and storage of hazardous chemicals, e.g. halogens, becoming increasingly difficult
- Alternatives for chlorinated hydrocarbon solvents, e.g. solvent-free processes, *chemistry in water*
- Shorter routes, in some cases via alternate feedstocks, e.g. substitution of alkanes for alkenes and aromatics
- Utilization, where feasible, of renewable raw-materials, e.g. carbohydrates, or waste materials from other processes
- Higher chemo-, regio- and enantioselectivities

Fig. 1. General trends in the chemical industry

2 Atom Utilization and Low Salt Technologies

Integrated waste management and zero emission plants are commonly heard phrases in the chemical corridors of power these days. What goes in must come out, somewhere. Preferably what comes out should be the desired product. Everything else should be considered as undesirable and its formation avoided or be kept in the system.

The most elegant solution is to avoid the formation of everything but the product by maximizing the *atom utilization*. The latter is calculated by dividing the molecular weight of the desired product by the sum total of all the materials (excluding solvents) used.

In traditional non-catalytic processes inorganic salts represent a large proportion of the total weight of material formed. Increasingly stringent environmental measures are stimulating the replacement of such traditional salt-forming technologies with cleaner catalytic alternatives (i.e. low-salt technologies).

An example, to illustrate the point, is the replacement of the traditional chlorohydrin technology of olefin epoxidation by catalytic epoxidation:

$$CH_2 = CH_2 + Cl_2 + H_2O \longrightarrow ClCH_2CH_2OH + HCl \qquad (1)$$

$$ClCH_2CH_2OH + HCl + Ca(OH)_2$$
$$\longrightarrow C_2H_4O + CaCl_2 + 2 H_2O \qquad (2)$$

atom utilization = 44/173 = 25%

$$CH_2 = CH_2 + 1/2 O_2 \xrightarrow{\text{catalyst}} C_2H_4O \qquad (3)$$

atom utilization = 100%

Similarly, catalytic oxidation of β-phenoxyethanols provides an interesting salt-free alternative to the classical process for the production of α-phenoxyacetic acid herbicides:

Classical process:

$$CH_3CO_2H + Cl_2 \longrightarrow ClCH_2CO_2H + HCl \qquad (4)$$

$$ClCH_2CO_2H + ArOH$$
$$\xrightarrow[\text{2. HCl}]{\text{1. 2 NaOH}} ArOCH_2CO_2H + 2 NaCl + 2 H_2O \qquad (5)$$

Overall stoichiometry:

$$ArOH + CH_3CO_2H + Cl_2 + 2 NaOH$$
$$\longrightarrow ArOCH_2CO_2H + 2 NaCl + 2 H_2O \qquad (6)$$

atom utilization* = 200/353 = 57%

*assuming Ar = 125

Low-salt alternative:

$$C_2H_4O + ArOH \xrightarrow{\text{H}^+} ArOCH_2CH_2OH \qquad (7)$$

$$ArOCH_2CH_2OH + O_2 \xrightarrow{\text{catalyst}} ArOCH_2CO_2H + H_2O \qquad (8)$$

atom utilization = 200/218 = 92%

Two technologies that are particularly clean in this respect are catalytic oxidation and catalytic carbonylation, since in principle they produce no salts as coproducts, e.g.

$$ArH + H_2O_2 \xrightarrow{\text{catalyst}} ArOH + H_2O \tag{9}$$

$$ROH + CO \xrightarrow{\text{catalyst}} RCO_2H \tag{10}$$

3 Bulk vs Fine Chemicals Manufacture

In the bulk chemicals industry classical environmentally unacceptable processes have largely (but not completely) been supplanted by cleaner, catalytic alternatives. Indeed, catalytic oxidation [1] is the single most important technology for the conversion of hydrocarbon feedstocks (alkanes, alkenes and aromatics) to industrial chemicals (see Table 1).

In the fine chemicals industry, in contrast, catalytic technologies have been only sporadically applied. One reason for this is the more or less separate development of the bulk and fine chemicals industries (see Fig. 2). The latter remained largely the domain of the synthetic organic chemist who, in general, did not apply catalytic methods. A second reason is the fact that absolute quantities

Table 1. Catalytic oxidation products

Product	Primary Raw materials	Volume[1] (10^6 tons)	Oxidant	Process[2]
Terephthalic acid	p-Xylene	4.0	O_2	L
Styrene	Benzene/ethylene	4.0	none	G
Formaldehyde	Methanol	3.0	O_2	G
Ethylene oxide	Ethylene	2.8	O_2	G
Phenol	a. Benzene/propylene b. Toluene	1.6	O_2	L
Acetic acid[3]	Ethylene/n-butane	1.6	O_2	L
Propylene oxide	Propylene	1.3	RO_2H	L
Acrylonitrile	Propylene	1.3	O_2/NH_3	G
Vinyl acetate	Ethylene	1.2	O_2	L/G
Acetone	Propylene	1.1	O_2	L/G
Benzoic acid	Toluene	1.0	O_2	L
Adipic acid	Benzene	0.9	O_2	L
Phthalic anhydride	o-Xylene	0.7	O_2	G
Methylmethacrylate	Isobutene	0.5	O_2	G
Acrylic acid	Propylene	0.5	O_2	G
Methyl ethyl ketone	1-Butene	0.3	O_2	G/L
Maleic anhydride	n-Butane	0.25	O_2	G

[1] U.S.A., 1989; [2] L = Liquid phase; G = Gas phase; [3] Acetic acid predominantly made by methanol carbonylation

25

R. A. Sheldon

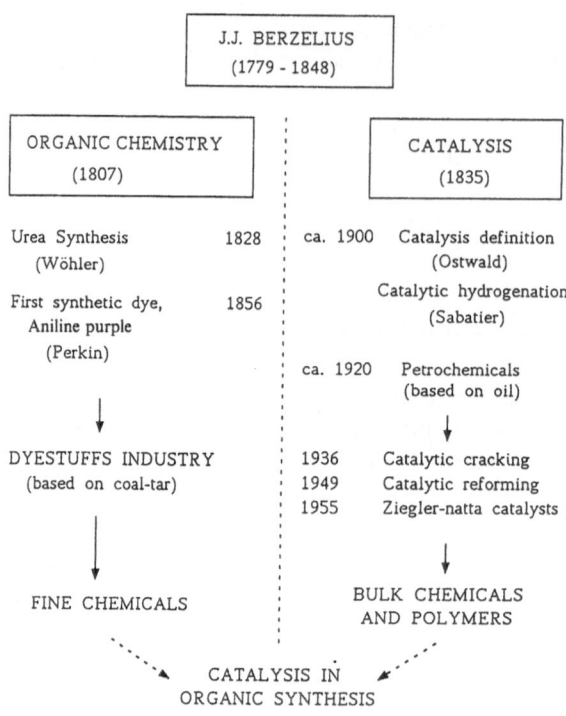

Fig. 2. The development of the bulk and fine chemicals industries

are significantly less in fine chemicals. Nevertheless, the number of kilos of byproducts (largely inorganic salts) per kilo of product are much higher in fine chemicals and specialties as indicated in Table 2.

The figures in Table 2 are gleaned from the author's own experience in bulk and fine chemicals and are intended to give an indication only. The increase on going downstream is partly due to the fact that fine chemicals and specialties are often produced in multi-step syntheses.

The fine chemicals industry, with its roots in coal-tar chemistry, positively abounds with processes involving classical, "stoichiometric" technologies, e.g. sulfonation, nitration, chlorination, bromination, diazotization, Friedel-Crafts

Table 2. Environmental Acceptability: The E Factor

Industry Segment	Product Tonnage	KG By-Products/ KG Product
Oil refining	$10^6 – 10^8$	ca. 0.1
Bulk chemicals	$10^4 – 10^6$	$<1–5$
Fine chemicals	$10^2 – 10^4$	$5–50$
Pharmaceuticals	$10 – 10^3$	$25– >100$

alkylations and acylations and stoichiometric oxidations, that produce large quantities of inorganic salt-containing aqueous effluent. Many of these processes are ripe for substitution by catalytic, low-salt technologies.

4 Characteristics of Fine Chemicals Manufacture and Choice of Oxidant

Although they share many common principles there are several basic differences between fine and bulk chemicals manufacture that can have an important bearing on process selection. Thus, fine chemicals are generally complex, multifunctional molecules which means that *chemo-*, *regio-* and *stereoselectivity* are important considerations. Such complex molecules usually have high boiling points and limited thermal stability, thus necessitating *reaction in the liquid phase* at moderate temperatures. Furthermore, processing tends to be *multi-purpose and batch-wise* in contrast to dedicated and continuous in-bulk chemicals. This means that not only raw materials costs but also *simplicity of operation and multi-purpose character of the installations* are important economic considerations (i.e. different ratio of variable to fixed costs).

Whereas in bulk chemical manufacture the choice of oxidant is largely restricted to molecular oxygen, the economics of fine chemicals production allow for a broader choice of oxidant (see Table 3). Indeed, even though it is more expensive per kilo than molecular oxygen, hydrogen peroxide is often the oxidant of choice for fine chemicals because of the simplicity of operation (i.e. lower fixed costs). Next to price and simplicity of handling the two important factors that influence the choice of oxidant are the nature of the coproduct and the percentage available oxygen. The former is obviously important in an environmental context and the latter directly influences the productivity (kg product per unit reactor volume per

Table 3. Oxygen donors

Donor	% Active oxygen	Coproduct
H_2O_2	47.0[1]	H_2O
O_3	33.3	O_2
CH_3CO_3H	26.6	CH_3CO_2H
$t\text{-BuO}_2H$	17.8	$t\text{-BuOH}$
NaClO	21.6	NaCl
$NaClO_2$	19.9[2]	NaCl
NaBrO	13.4	NaBr
HNO_3	25.4	NO_x
$C_5H_{11}NO_2^3$	13.7	$C_5H_{11}NO$
$KHSO_5$	10.5	$KHSO_4$
$NaIO_4$	7.0_2	NaI
PhIO	7.3	PhI

[1] Based on 100% H_2O_2
[2] Assuming that only one oxygen atom is utilized
[3] *N*-Methylmorpholine-N-oxide

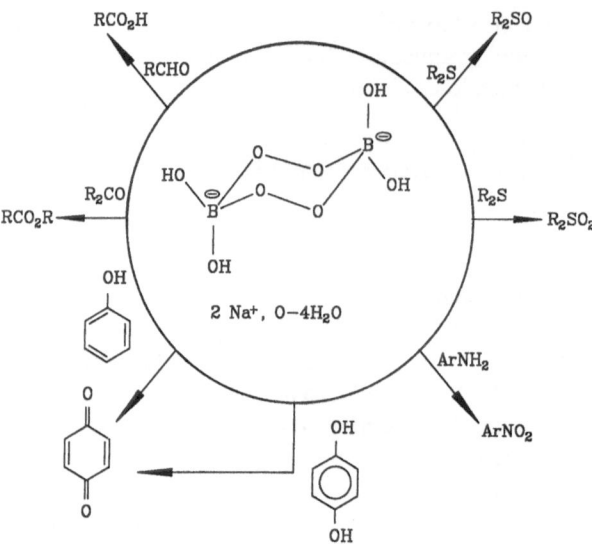

Fig. 3. Oxidations with sodium perborate. Current Production: $> 10^6$ t year^{-1} Major Uses: Detergents, Antiseptic

unit time). The percentage of available oxygen is also simply another way of expressing the atom utilization of the oxidant in question. Hydrogen peroxide is obviously "Mr. Clean", its coproduct being water. It should be noted, however, that the coproduct from organic oxidants, e.g. *tert*-butylhydroperoxide (TBHP), peracetic acid, and amine oxides, can be readily recycled via reaction with hydrogen peroxide. The overall process affords water as the coproduct, but requires one extra step compared to the corresponding reactions with H_2O_2. Thus, peroxide reagents are potentially ideal oxygen transfer reagents.

With inorganic oxygen donors environmental considerations are relative. Thus, sodium chloride (from NaClO or NaClO$_2$) and potassium sulfate (from KHSO$_5$) are obviously preferred to chromium, manganese or lead salts. Sodium bromide would seem, in the first instance, to be less attractive than sodium chloride. On the other hand, bromine can be regenerated fairly simply by reaction of bromide with H_2O_2, providing an overall recycle process with water as the coproduct. In addition to the reagents listed in Table 3 there are other potentially useful oxygen donors. For example, sodium perborate is produced on more than a million tons per annum scale for use in detergents and antiseptic mouthwashes. McKillop and coworkers[2] have reported its use, in the absence of metal catalysts, as a selective oxidant in organic synthesis (see Fig. 3).

5 Catalytic Oxidation vs Oxygen Transfer

The overwhelming majority of petrochemical oxidation processes involve either free radical autoxidation in the liquid phase or gas phase oxidation over solid catalysts. The former type of process displays rather indiscriminate reactivity and is selective only with substrates containing one group that is susceptible to radical attack (e.g. cumene autoxidation and oxidation of toluene or *p*-xylene to benzoic and terephthalic acid, respectively). The latter type of process often involves a Mars-van-Krevelen mechanism [2], i.e. stoichiometric oxidation of the substrate followed by re-oxidation of the reduced form of the oxidant with molecular oxygen (see Fig. 4). In general this type of process is successful only with relatively simple, volatile molecules, e.g. maleic anhydride from *n*-butane, phthalic anhydride from *o*-xylene and acrylonitrile by propylene ammoxidation.

LIQUID PHASE:

$$RH \xrightarrow{\text{initiation}} R^{\bullet}$$

$$R^{\bullet} + O_2 \longrightarrow RO_2^{\bullet}$$

$$RO_2^{\bullet} + RH \longrightarrow RO_2H + R^{\bullet}$$

GAS PHASE (MARS-VAN-KREVELEN MECHANISM):

$$S + M = O \longrightarrow SO + M$$

$$2M + O_2 \longrightarrow 2M = O$$

S = substrate
M = O is the oxidized form of the catalyst (e.g. V_2O_5)

Fig. 4. Mechanisms for oxidation with O_2 in the liquid and gas phase

In the gas phase a Mars-van-Krevelen type mechanism can compete successfully with free radical autoxidation because if free radicals are formed they are not surrounded by substrate molecules (RH) as in the liquid phase, i.e. free radical chains are very short. Conversely, a Mars-van-Krevelen type mechanism is difficult to achieve with O_2 in the liquid phase (but see later) due to competing facile autoxidation. The key to designing selective catalysts for liquid phase oxidations is to create a "gas phase environment" in the liquid phase.

Thus, in general catalytic oxidations with molecular oxygen are not applicable to more complicated, less volatile molecules. For reactions in the liquid phase *catalytic oxygen transfer* constitutes a useful alternative that combines the advantages of stoichiometric oxidants (high selectivities and broad scope) with those of catalytic oxidation with O_2 (inexpensive reagent and environmentally acceptable) [3, 4]. A reaction of commercial interest, to illustrate the point, is the oxidation of 2-methylnaphthalene to 2-methyl-1,4-naphthoquinone (menadione), an intermediate for vitamin K. Traditionally this oxidation was carried out with stoichiometric quantities of chromium trioxide in sulfuric acid and produced 18 kg of inorganic salts per kg of product. An alternative procedure, employing 60% aqueous H_2O_2 and a palladium(II)-exchanged polystyrene sulfonic acid resin as catalyst (reaction 11) has been reported [5].

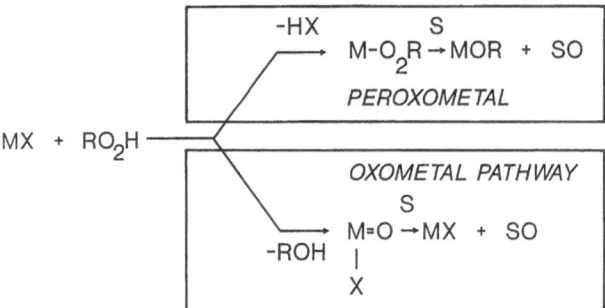

(11)

(55–60% yield)

Oxygen transfer involves reaction of an oxygen donor (see Table 3) with an organic substrate in the presence of a metal catalyst:

$$M + XOY \longrightarrow \text{ACTIVE OXIDANT} \xrightarrow{S} M + SO + XY$$

$$XOY = \text{oxygen donor}; \quad S = \text{substrate}; \quad M = \text{catalyst}$$

The active oxidant in these processes may be an oxometal or a peroxometal species (see Fig. 5). Some metals (e.g. vanadium) can, depending on the substrate, operate via either mechanism.

$$MX + RO_2H$$

-HX $\quad$ S

$M-O_2R \rightarrow MOR + SO$

PEROXOMETAL

OXOMETAL PATHWAY

S

$M=O \rightarrow MX + SO$

-ROH $\quad |$

$\quad$ X

Fig. 5. Peroxometal vs oxometal pathways

6 Examples of Catalytic Oxygen Transfer

Virtually all of the transition elements and several main group elements (e.g. Sn, As, Se) can be used in combination with a variety of relatively inexpensive oxygen donors, providing for an enormous number of possible combinations.

The most well-known example is the catalytic epoxidation of olefins with alkyl hydroperoxides that is used for the commercial production of propylene oxide (see earlier). The reaction is catalyzed by high-valent compounds of early transition metals, e.g. Mo^{VI}, W^{VI}, V^V and Ti^{IV}, and involves a peroxometal type mechanism [6,7] as shown (reaction 12). Mo compounds are particularly effective homo-

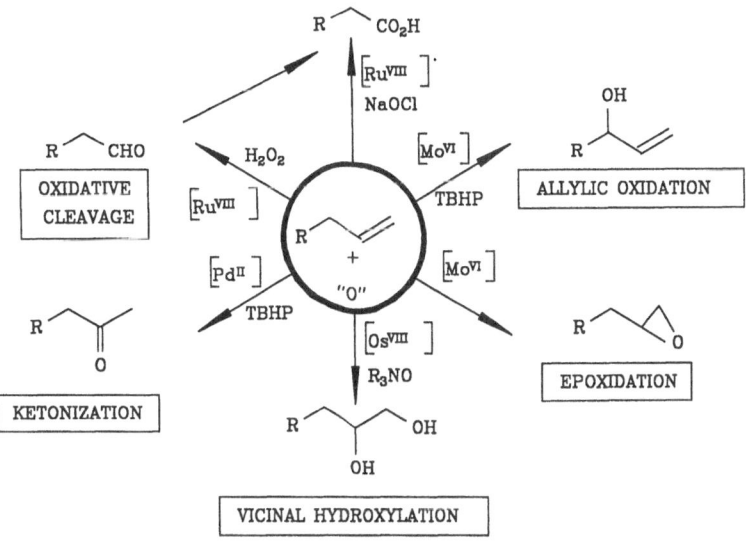

Fig. 6. Oxidative transformations of olefins

geneous catalysts and a Ti^{IV}/SiO_2 catalyst developed by Shell is a highly effective heterogeneous catalyst for this reaction. Analogous chemo-, regio- and stereo-selective epoxidations can be carried out with these RO_2H-metal catalyst reagents under relatively mild conditions (hydrocarbon solvent; 80–120 °C).

$$\tag{12}$$

The versatility of metal catalyst-oxygen donor reagents in organic synthesis is further illustrated by the variety of other useful oxidative transformations of olefins [11] that they are able to effect (Fig. 6).

Probably the most frequently encountered oxidative transformation in organic synthesis is alcohol oxidation. Traditionally, such transformations were (and are) carried out using stoichiometric, chromium(VI)-based reagents. Although very popular among organic chemists such reagents are not compatible with current "environmentality". There is good news, however, for synthetic organic chemists: a variety of environmentally attractive catalytic oxygen transfer reagents are available for the chemo- and regioselective oxidation of alcohols and diols [3]. A few examples are shown below.

$$\xrightarrow[\text{VO(acac)}_2]{\text{TBHP}}$$

96% yield

[Ref. 8] $\qquad$ (13)

$$\text{Ph}\diagup\!\!\!\diagup\!\!\overset{\overset{\displaystyle OH}{|}}{\diagdown}\xrightarrow[\left[Cr^{III}/\text{NAFK}\right]]{\text{TBHP}} \text{Ph}\diagup\!\!\!\diagup\!\!\overset{\overset{\displaystyle O}{\|}}{\diagdown}\qquad\text{[Ref. 9]}\qquad(14)$$

81% yield

$$\xrightarrow[\left[Ce^{IV}/\text{NAFK}\right]]{\text{TBHP}}\qquad\text{[Ref. 9]}\qquad(15)$$

98% yield

$$\text{RCH(OH)CN}\xrightarrow[\left[\text{RuCl}_2\text{L}_2\right]]{\text{TBHP}}\underset{\overset{\|}{O}}{\text{RCCN}}\qquad\text{[Ref. 10]}\qquad(16)$$

72–99% yield

Ruthenium compounds are the catalysts par excellence for the chemoselective oxidations of alcohols and diols and can be used in conjunction with a wide variety of oxygen donors [11], such as H_2O_2, NaOCl, RO_2H, N-methylmorpholine oxide (NMO). Alcohol oxidations can involve peroxometal or oxometal pathways depending on the catalyst used. Thus, metals that are strong oxidizing agents in high oxidation states (e.g. Ru^{VIII}, Cr^{VI}, V^V, Ce^{IV}) react via oxometal pathways whilst weakly oxidizing metal ions (e.g. Mo^{VI}, W^{VI}, Zr^{IV}, Ti^{IV}) involve peroxometal species in the key oxidative dehydrogenation step.

Another reaction of practical interest is the oxidative cleavage of vicinal diols (reaction 17) that is traditionally achieved using the stoichiometric oxidants, periodates or lead tetraacetate.

$$\overset{\overset{\displaystyle OH\quad OH}{|\quad\ \ |}}{\diagup\!\!\overset{}{C}\!-\!\overset{}{C}\!\diagdown}\xrightarrow[\left[\text{catalyst}\right]]{\text{OXYGEN DONOR}}\diagup\!\!C\!=\!O\ +\ O\!=\!C\!\diagdown\qquad(17)$$

A variety of the oxygen transfer reagents described above, e.g. VO(acac)$_2$/TBHP [12], W^{VI}/H_2O_2 [13, 14], RuCl$_3$/H$_2$O$_2$ [15] and RuCl$_3$/NaOCl [16] have been successfully applied to this transformation, thus providing attractive alternatives to the classical stoichiometric oxidants. An interesting variation on this theme is the use of ruthenium pyrochlore oxides ($A_{2+x}Ru_{2-x}O_{7-y}$ where A is Pb or Bi) as heterogeneous catalysts for the liquid phase oxidative cleavage of vicinal diols by molecular oxygen under relatively mild conditions, e.g. in the conversion of cyclohexane-1,2-diol to adipic acid [17]:

$$\xrightarrow[\text{catalyst}]{O_2,\ \text{Aq.NaOH}}\qquad(18)$$

81–87% yield

Catalyst; Pb$_x$Ru$_y$O$_x$ or Bi$_x$Ru$_y$O$_x$

We were recently interested in the selective cleavage of carbohydrates as a method for producing commercially interesting chiral C_3 synthons, e.g. as shown in reaction 25. We found [18] that both homogeneous and heterogeneous ruthenium catalysts were effective in conjunction with NaOCl as the oxygen donor. In this case the ruthenium pyrochlore catalyst worked well with NaOCl but not with molecular oxygen. Reaction 19 is interesting for several reasons: it utilizes a "green" raw material and a catalytic process to produce a single enantiomer, i.e. a product devoid of isomeric ballast.

GREEN RAW MATERIALS: CHIRONS
FROM CATALYTIC CONVERSION OF CARBOHYDRATES

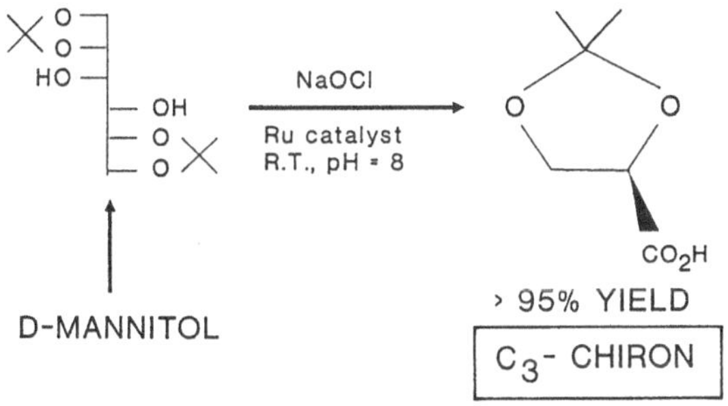

(19)

> 95% YIELD

C_3- CHIRON

D-MANNITOL

- HOMOGENEOUS CATALYST: RuCl$_3$ (1 - 5 mol %)

- HETEROGENEOUS CATALYST: Ru/C or Ru PYROCHLORE OXIDE
(2.5 mol %)

Ruthenium is certainly the most versatile catalyst in oxygen transfer processes. The recently reported [19] highly selective oxidations of beta-lactams shown in Fig. 7 illustrate what is possible with these reagents.

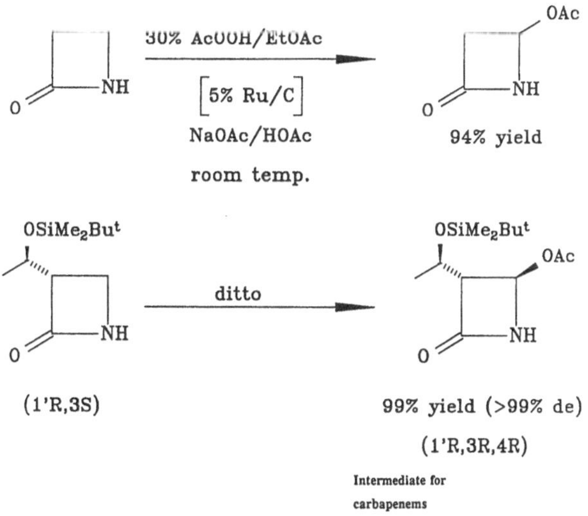

Fig. 7. Catalytic oxidation of beta-lactams

7 Ligand-Mediated Catalytic Oxidations

In the examples discussed so far, the ligands surrounding the metal do not play a crucial role in the observed reaction. Indeed, it is generally not shown what the ligands are. However, the ligands surrounding the metal can, in principle, influence the *activity*, *selectivity* and *stability* of a catalyst, i.e. we can speak of *ligand-mediated metal catalysis*. Good examples are provided by processes catalyzed by metal-loenzymes where not only the metal but also the ligand play an essential role in the reaction. In catalytic oxidations, however, there is one serious problem: most organic ligands are thermodynamically unstable in oxidizing media.

This problem is well illustrated by the ubiquitous cytochrome P450-dependent mono-oxygenases [20], the prosthetic group of which contains an iron(III)porphyrin complex. The porphyrin ligand is essential for the reactivity of the catalyst. It promotes the reaction with molecular oxygen and it stabilizes the high-valent oxoiron porphyrin species that constitutes the active oxidant in these sytems (see Fig. 8). Cytochrome P450-dependent mono-oxygenases can also be operated in an oxygen transfer mode whereby the active oxoiron species is formed directly by reaction of the iron(III) porphyrin with an oxygen donor (the so-called peroxide shunt).

However, this powerful oxidant is not only capable of oxidizing a wide variety of organic substrates but it also self-destructs by oxidative degradation of its own porphyrin ligand. This means that cytochrome P450-containing enzymes are not stable for any length of time outside the cell, which is a serious drawback in practice. Consequently, much effort has been devoted to the design of simple model systems [21] capable of effecting the same, often highly regio- and stereoselective

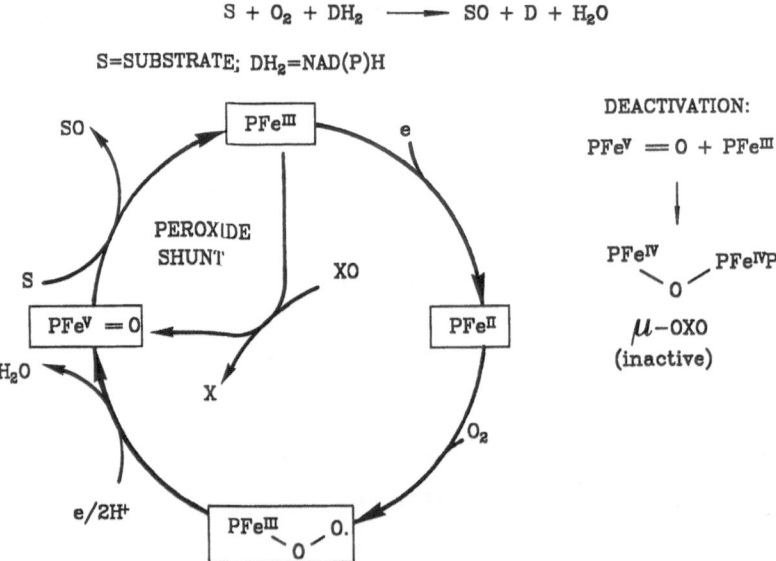

Fig. 8. The catalytic cycle of cytochrome-P450

oxidations. Most of these model systems consist of iron or manganese porphyrin catalysts in combination with single oxygen donors such as NaOCl [22], $KHSO_5$ [23] and in a few cases H_2O_2 [24]. Unfortunately, virtually all of these model systems suffer from the same disadvantage as the enzyme that they attempt to emulate, i.e. they contain expensive, unstable ligands.

A second disadvantage encountered with these biomimetic model systems, and with many other oxidations involving oxometal species, is deactivation via the formation of unreactive dimeric (or oligomeric) μ-oxo complexes, e.g.

$$PM^V = O + PM^{III} \longrightarrow PM^{IV} - O - M^{IV}P \tag{20}$$
$$\mu\text{-oxo}$$

$$M = Mn, Fe, etc.$$

$$Ti^{IV} = O \longrightarrow \{Ti^{IV} - O - Ti^{IV} - O\}_n \tag{21}$$

There remains, therefore, a need for the development of ligands that can not only stabilize high-valent oxometal species but are also stable with respect to both oxidative self-destruction and deactivation via μ-oxo complex formation (N.B. the cyt-P450 oxoiron species cannot form a μ-oxo dimer for steric reasons). One promising approach involves the use of stable polyhalogenated porphyrin ligands [25, 26]. For example, Ellis and Lyons [25] have recently reported that iron(III) complexes with the highly stable perhalogenated porphyrin ligands catalyze the unprecedented selective hydroxylation of isobutane with molecular oxygen at ambient temperatures. In order to explain their high activity it was postulated [25] that polyhalogenation of the porphyrin ring system stabilizes the active oxoiron intermediate with respect to μ-oxo dimer formation. Obviously such oxidatively stable, highly active catalysts should havse broad utility for selective oxidation and it would be interesting to use them in combination with single oxygen donors such as H_2O_2.

8 Site-Isolation of Oxometal Species in Inorganic Matrices — Mineral Enzymes

Another promising approach to constructing stable, active oxidation catalysts is to isolate the appropriate high-valent metal species in a stable inorganic matrix such as a heteropoly acid (anion) or in the lattice of a zeolite or related microporous solids (e.g. alpos, silicalite, etc.). Indeed, oxidic species may be classified according to their degree of aggregation, increasing in complexity from monomeric soluble oxometal complexes, through oligomeric heteropoly acids, to polymeric zeolites, pillared clays etc.

9 Heteropolyacids and Phase Transfer Catalysis

Heteropolyacids (HPA's) [27] and their salts are polyoxo compounds incorporating anions (heteropolyanions) having metal-oxygen octahedra (MO^6) as the basic structural unit. They contain one or more heteroatoms (Si, Ge, P, As, etc.) that are usually located at the center of the anion. The MO_6 octahedra are linked together to form an extremely stable and compact structure for the heteropoly-anion. One of the most common types of HPA comprises the so-called Keggin anions, $XM_n^1 M_{12-n}^2 O_{40}^X$ (where $M^1 = Mo^{VI}$, W^{VI} and $M^2 = V^V$). Despite their rather awesome formulae HPA's are easy to synthesize by acidification of aqueous solutions containing the heteroelement and the appropriate mixture of alkali metal molybdate, tungstate and vanadate. They possess several rather unique properties that render them potentially interesting oxidation catalysts. Thus, they are both strong Brönsted acids and multi-electron oxidants, i.e. they are potential bifunc-tional catalysts. They are, moreover, soluble in water and oxygenated organic solvents which means they can be regarded as "soluble oxides".

Molybdenum and tungsten heteropolyanions with the general formula $H_3PM_{12}O_{40}$ (M = Mo or W) in conjunction, with the phase transfer agent cetylpyridinium chloride, have been widely applied by Ishii and coworkers [14, 28] in the epoxidation of olefins, oxidation of allylic alcohols and oxidative cleavage of olefins and vicinal diols with H_2O_2 in one- or two-phase systems. The oxidative cleavage reactions, for example, gave the best results under homogeneous conditions in *tert*-butanol as solvent, e.g.

$$(22)$$

The scope of HPA's as selective oxidation catalysts has been further extended by incorporating other redox metal ions, whereby the heteropolyanion can be regarded as an oxidatively resistant inorganic analogue of the porphyrin ligand. For example, manganese (II) and cobalt(II)-substituted heteropolytungstates of general formula $(R_4N)_4HMPW_{11}O_{39}$ (M = Mn^{II} or Co^{II}) catalyze the epoxidation of olefins with PhIO and the hydroxylation of alkanes with TBHP, in complete analogy to the cytochrome P450 model systems described above [29]. We conclude, on the basis of the examples outlined above, that heteropolyacid-based catalysts have a promising future as selective oxidation catalysts in organic synthesis.

It should be noted, however, that although heteropolyacids are stable to oxidative degradation they are often not completely stable to dissociation to smaller units, especially in protic solvents. Moreover, they occupy an intermediate position between homogeneous metal complexes and truly heterogeneous oxidic catalysts.

10 Redox Zeolites and Pillared Clays

A second approach to isolating redox metal ions in stable inorganic matrices, thereby creating oxidation catalysts with interesting activities and selectivities, is to incorporate them in a zeolite lattice framework. The first example of such a "redox zeolite" was the synthetic titanium(IV) zeolite, titanium silicalite (TS-1), developed by Enichem [30–32]. TS-1 was shown to catalyze a variety of synthetically useful oxidations with 30% H_2O_2, such as olefin epoxidation, oxidation of primary alcohols to aldehydes, aromatic hydroxylation, and ammoximation of cyclohexanone to cyclohexanone oxime (see Fig. 9).

The TS-1 catalyst exhibits some quite remarkable activities and selectivities, e.g. ethylene, is epoxidized with 30% H_2O_2 in aqueous *tert*-butanol at ambient temperature, affording ethylene oxide in 96% selectivity at 97% H_2O_2 conversion.

The TS-1 catalyzed hydroxylation of phenol to a 1:1 mixture of catechol and hydroquinone has already been commercialized by Enichem. Another reaction of considerable commercial importance is the ammoximation of cyclohexanone to cyclohexanone oxime, an intermediate in the manufacture of caprolactam. It could form an attractive alternative to the established process that involves a circuitous route via oxidation of ammonia to nitric acid followed by reduction of the latter to hydroxylamine (see Fig. 10). The ammoximation route employs a more expensive oxidant (H_2O_2) but is shorter and produces considerably less salt. However, we note that is does not provide a complete solution to the salt problem as substantial amounts are also produced in the subsequent Beckmann rearrangement of the oxime. The answer to this problem is probably also in the deployment of an efficient solid catalyst.

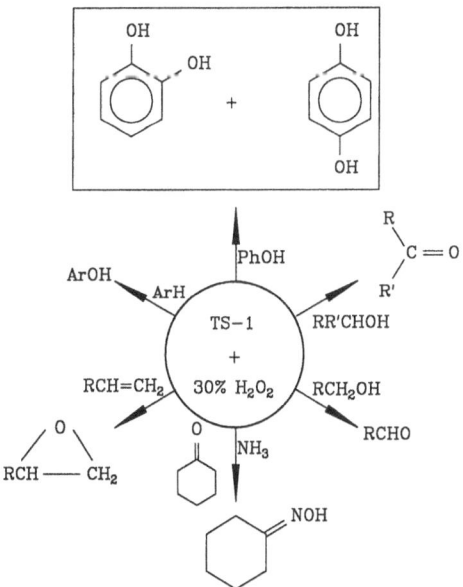

Fig. 9. Oxidation catalyzed by titanium silicalite (TS-1)

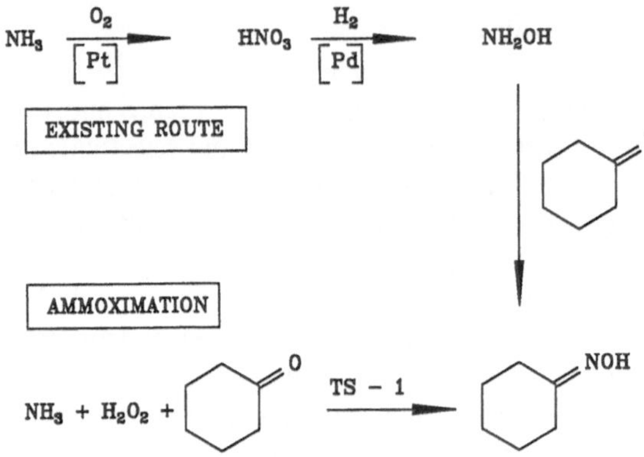

Fig. 10. Two routes to cyclohexanone oxime

11 Shape Selective Oxidation

Incorporating redox catalytic sites within a zeolite lattice framework should also provide a basis for effecting shape selective oxidations. Indeed, it has recently been reported [33–35] that TS-1 catalyzes the shape selective oxidation of alkanes with 30% H_2O_2. Linear alkanes were oxidized much faster than branched or cyclic alkanes, presumably as a result of the molecular sieving action of TS-1. The products were the corresponding alcohols and ketones formed by oxidation at the 2- and 3-positions, e.g.

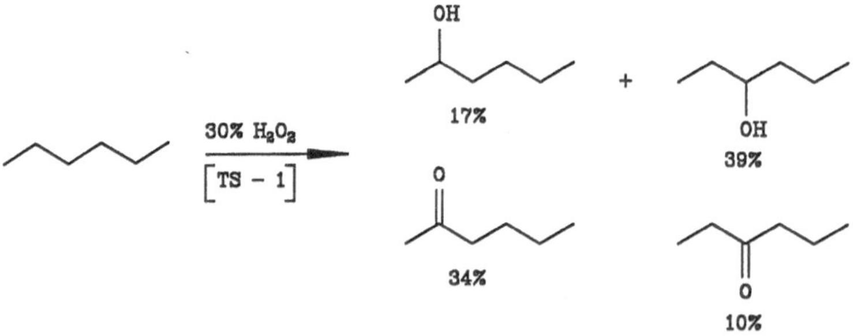

Interestingly, oxidation of *n*-hexane with TS-1/H_2O_2 in methanol solvent was faster than the corresponding epoxidation of 1-octene. In competition experiments, in contrast, 1-octene reacted 40 times as fast as *n*-hexane, indicating that the olefin preferentially complexes to the catalyst [36].

From a mechanistic viewpoint it is worth noting that the TS-1 catalyst contains the same chemical elements in roughly the same proportions as the Shell amorphous Ti^{IV}/SiO_2 catalyst referred to earlier. The question arises as to what is the explanation for the vastly different reactivities of the two catalysts, e.g. the Shell catalyst is inactive with H_2O_2 as the oxidant. A possible explanation for the high reactivity of TS-1 is that the hydrophobic cavity containing the active peroxotitanium(IV) oxidant is only large enough (5–6 Å) to accommodate the

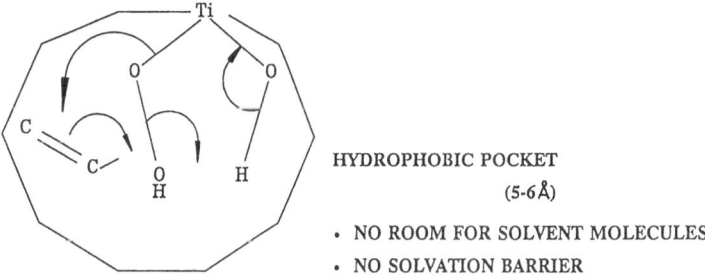

HYDROPHOBIC POCKET

(5-6Å)

- NO ROOM FOR SOLVENT MOLECULES
- NO SOLVATION BARRIER

Fig. 11. Catalysis inside TS-1: an artist's impression

substrate with little or no room for solvent and/or water molecules (see Fig. 11). This means that little or no solvation energy has to be overcome, i.e. the reaction can be compared to a reaction under gas phase conditions.

Based on the quite remarkable results obtained with TS-1 we expect that many more examples of microporous solids modified by isomorphous substitution with redox metal ions in the crystal lattice will be described in the future (see Fig. 12). Indeed, the scope for developing unique oxidation catalysts based on the concept of *site-isolation* in zeolites, silicalites, alpos and sapos is enormous [37]. In addition to varying the redox metal the size and hydrophobicity of the cavity can be tuned by, for example, varying Si/Al ratios to provide a variety of unique heterogeneous catalysts for liquid phase oxidation.

There is considerable current interest in the design of new catalysts by interchelating clay minerals of the smectite type with redox metal ions, leading to the formation of oxidation catalysts with interesting (shape-selective) properties [38]. For example, vanadium-pillared montmorillonite (V-PILC) proved to be an

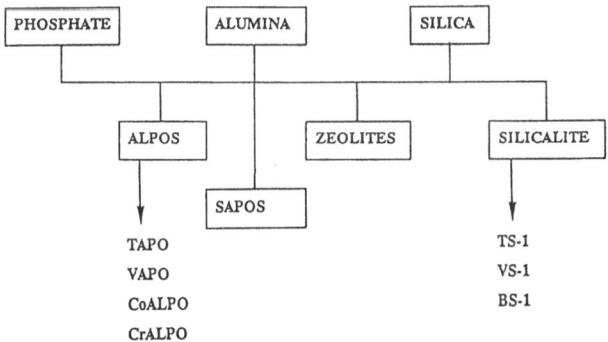

- ISOLATED (OXO)METAL CENTERS WITH UNIQUE ACTIVITIES & SELECTIVITIES
- SHAPE-SELECTIVE POCKETS WITH TUNABLE HYDROPHOBIC/HYDROPHILIC CHARACTER i.e. MINERAL ENZYMES

Fig. 12. Types of molecular sieves

effective, shape-selective catalyst for the oxidation of benzylic alcohols to the corresponding benzoic acids, using 30% aqueous H_2O_2 as the terminal oxidant [38]. Thus p-substituted benzyl alcohols were oxidized whilst o-substituted benzyl alcohols were essentially inert [38].

Similarly, chromia-pillared montmorillonite catalyst reportedly [39] catalyzes the selective oxidation of alcohols with TBHP and titanium-montmorillonite in conjunction with tartaric acid esters functions as a heterogeneous catalyst for asymmetric epoxidations with TBHP [40].

12 Catalysis and Chirality — Enantioselective Oxidation

One of the most challenging goals in catalysis research is the design of simple chemocatalysts that can achieve the high levels of enantioselectivity that are characteristic of many biocatalytic oxidations. An important prerequisite for high enantioselectivity in such processes is that coordination of the chiral ligand to the metal ion results in a substantial rate acceleration. Sharpless [41] coined the term *ligand accelerated catalysis* to describe this phenomenon. Thus, if the metal-chiral ligand complex rapidly exchanges its ligands in solution the high enantioselectivities will only be observed when M–L is a much more active catalyst than M (see reaction 24).

$$M + L_{chiral} \longrightarrow M-L_{chiral} \tag{24}$$
$$\text{achiral catalyst} \qquad \text{chiral catalyst}$$

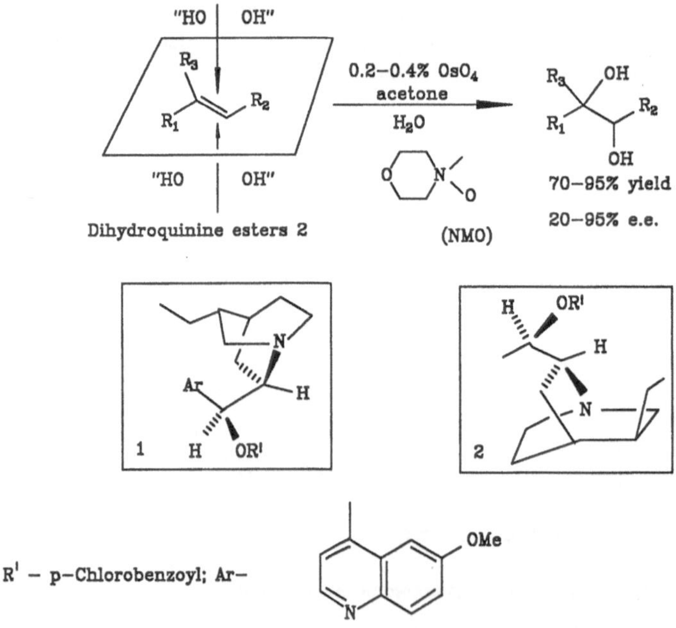

Fig. 13. Ligand accelerated asymmetric catalysis

This situation obtains for example in the asymmetric vicinal dihydroxylation of olefins (see Fig. 13) reported by Sharpless and coworkers [41]. Coordination of a (chiral) amine to the OsO_4 catalyst affords a (chiral) catalyst with much higher activity.

13 Enzymes as Catalysts

The best examples of ligand accelerated catalysis are, of course, enzymatic processes where the protein ligands are responsible for enormous rate accelerations. So why not use enzymes themselves as catalysts? As noted earlier, with monooxygenases there is often a problem of stability and there is the question of cofactor requirement. A more promising approach therefore is to use peroxidases that are more robust than the monooxygenases and are able to utilize H_2O_2. It is interesting in this context, and in that of enzymatic processes in general, to use enzymes under conditions and with substrates that nature did not design them for. Why should enzymes be limited to the tasks that they were designed for in nature? A good example of what can be achieved is the recently reported [42] enantioselective sulfoxidation with H_2O_2 in the presence of chloroperoxidase (Fig. 14). The latter was definitely not designed for carrying out this reaction.

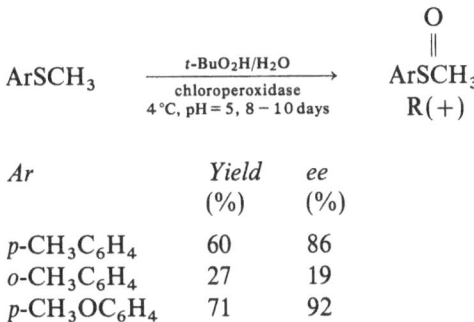

Ar	Yield (%)	ee (%)
p-$CH_3C_6H_4$	60	86
o-$CH_3C_6H_4$	27	19
p-$CH_3OC_6H_4$	71	92

Fig. 14. Enantioselective sulfoxidation

Another example where an enzyme is used to mediate a completely different reaction than it normally does is the lipase-catalyzed formation of peracids by reaction of H_2O_2 with a carboxylic acid [43]. As illustrated in Fig. 15 this allows for the one step epoxidation of olefins by in situ formation of RCO_3H. The carboxylic acid can be used in catalytic amounts providing an overall epoxidation with H_2O_2. By a suitable choice of carboxylic acid the reaction can be carried out in a two-phase system. The scope of such novel transformations must be enormous.

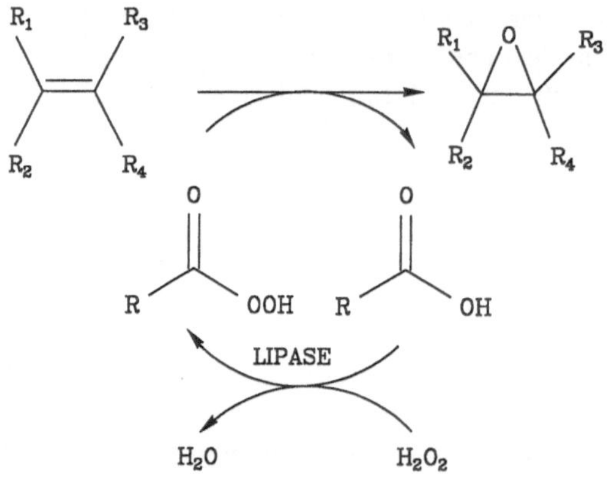

R. A. Sheldon

Fig. 15. Lipase-catalyzed, in-situ formation of peracids

14 Conclusions and Future Prospects

Based on the above discussion we conclude that catalytic oxidation is an area in which many exciting new developments are taking place. Driven by the need for cleaner processes, especially in fine chemicals, and shorter routes with higher selectivities, many interesting new technologies are emerging. Catalytic oxygen transfer processes utilizing peroxide are particularly interesting in this context.

In particular we expect that the design of novel heterogeneous catalysts for liquid phase oxidations will play an important role in these developments. Not only because they have the inherent advantage of ease of recovery and recycling but also because they offer the possibility of designing site-isolated redox metal catalysts displaying unique substrate, chemo-, regio- and stereoselectivities. Furthermore, we expect that more and better methods for enantioselective oxidation will be developed, possibly using enzymes under unnatural conditions.

15 References

1. Sheldon RA, Kochi JK (1981) Metal-catalyzed oxidations of organic compounds, Academic, New York; Sheldon RA (1991) Stud Surf Sci Catal 66: 573; 33
2. McKillop A, Tarbin JA (1983) Tetrahedron Lett 24: 1505 (1987) Tetrahedron 43: 1753
3. Mars P and van Krevelen DW (1954) Chem Eng Sci Special Supplement, 3: 41
4. Sheldon RA (1990) Stud Surf Sci Catal 55: 1 and references cited therein; Sheldon RA (1985) Bull Soc Chim Belg 94: 651
5. Yamaguchi S, Inoue M, and Enomoto S (1985) Chem Lett 827
6. Sheldon RA (1981) In: Ugo R (ed) Aspects of homogeneous catalysis, vol 4, Reidel, Dordrecht, p 1
7. Jorgensen KA (1989) Chem Rev 89: 431
8. Kaneda K, Kawanishi Y, Jitsukawa K and Teranishi S (1983) Tetrahedron Lett 24: 5009
9. Kanemoto S, Saimoto H, Oshima K and Nozaki H (1984) Tetrahedron Lett 25: 3317
10. Tanaka M, Kobayashi T and Sakakura T (1984) Angew Chem 96: 519
11. Griffith WP (1989) Plat Metals Rev 33: 181

12. Zviely M, Goldman A, Kirson I and Glottes E (1986) J Chem Soc Perkin Trans 229
13. Venturello C and Ricci M (1986) J Org Chem 51: 1599
14. Yamawaki K, Yoshida T, Nishihara H, Ishii Y and Ogawa M (1986) Synth Commun 16: 537; Ishii Y, Yamawaki K, Ura T, Yamada H, Yoshida T and Ogawa M, J (1988) Org Chem 53: 3587
15. Japanese Patent 102 527 (1980) to Mitsui Petrochemical CA 94: 46783b (1981)
16. Wolfe S, Hasan SK and Campbell JR (1970) Chem Commun p 1420
17. Felthouse TR (1987) J Am Chem Soc 109: 7566
18. Emons C, Vekemans J, Kuster B and Sheldon RA (1991) Tetrahedron Asymmetry 2: 359
19. Murahashi S, Naota T, Kuwabara T, Saito T, Kumobayashi H and Akutagawa S (1990) J Am Chem Soc 112: 7820
20. Ortiz de Montellano P (ed) (1986) Cytochrome P450: Structure, mechanism and bio-chemistry, Plenum, New York
21. For leading references see: Meunier B (1988) Gazz Chim Ital 118: 485
22. Meunier B, De Carvalho ME and Robert A (1987) J Mol Catal 41: 185; van der Made AW, van Gerwen MJP, Drenth W and Nolte RJM (1987) J Chem Soc Chem Commun 888
23. De Poorter B, Ricci B and Meunier B, (1985) Tetrahedron Lett 26: 4459
24. Battioni P, Renaud JP, Bartoli JF, Reine-Artiles M, Fort M and Mansuy D (1988) J Am Chem Soc 110: 8462
25. Ellis PE and Lyons JE (1989) Catal Lett 3: 389; Lyons JE and Ellis PE (1991) ibid 8: 45
26. Anelli PL, Banfi S, Montanari M and Quici S (1989) J Chem Soc Chem Commun 779
27. For recent reviews see: Kozhenikov IV (1987) Russ Chem Rev 56: 811; Kozhenikov IV and Matveev KI (1983) Appl Catal 5: 135; Misono M (1987) Catal Rev Sci Eng 29: 269
28. Ishii Y in Reviews on Heteroatom Chemistry, vol 3, Oae S (ed) MYU, Tokyo 1990 pp 121
29. Hill CL in Activation and functionalization of alkanes, Hiu CL (ed) Wiley, New York, 1989, pp 243
30. Romano U, Esposito A, Maspero F, Neri C and Clerici MG (1990) Chim Ind (Milan) 72: 610 and references cited therein
31. Notari B (1988) Stud Surf Sci Catal 37: 413
32. Romano U, Esposito A, Maspero F and Clerici MG (1990) Stud Surf Sci Catal 55: 33
33. Tatsumi T, Nakamura M, Negishi S and Tominaga H (1990) J Chem Soc Chem Commun 476
34. Huybrechts DRC, De Bruyker L and Jacobs PA (1990) Nature, 345: 240
35. Clerici MG (1991) Appl Catal 68: 249
36. Huybrechts DRC, PhD thesis, Catholic University of Leuven, Belgium, 1991
37. Wilson ST in Introduction to zeolite science and practice, Van Bekkum H, Flanigen EM and Jansen JC (eds) Elsevier, Amsterdam, 1991
38. Choudary BM and Valli VLK (1990) J Chem Soc Chem Commun 721
39. Choudary BM, Durga Prasad A and Valli VLK (1990) Tetrahedron Lett 31: 5785
40. Choudary BM and Valli VLK (1990) J Chem Soc Chem Commun 1115
41. Jacobsen EN, Marko I, Mungall WS, Schröder G and Sharpless KB (1988) J Am Chem Soc 110: 1968; Jacobsen EN, Marko I, France MB, Svendsen JS and Sharpless KB (1989) ibid 111: 737; Wai JSM, Marko I, Svendsen JS, Finn MG, Jacobsen EN and Sharpless KB (1989) ibid 111: 1123; Lohray BB, Kalentar TH, Kim BM, Park CY, Shibata T, Wai JS and Sharpless KB (1989) Tetrahedron Lett 30: 2041
42. Colonna S, Gaggero N, Manfredi A, Casella L, Gullotti M, Carrea G and Pasta P (1990) 29: 10465
43. Björkling F, Godtfredsen SE and Krik O (1990) J Chem Soc Chem Commun 1301

Dioxiranes: Oxidation Chemistry Made Easy

Waldemar Adam*, Lazaros Hadjiarapoglou

Institute of Organic Chemistry, University of Würzburg, Am Hubland, D-8700 Würzburg, FRG

Table of Contents

1 General Aspects

Since its isolation in 1985 by Murray and Jeyaraman [2], dimethyldioxirane (DMD), as acetone solutions, has become a very important oxidant for preparative oxidation chemistry [3]. This novel three-membered ring cyclic peroxide constitutes an ideal oxidant in that it is efficient in its oxygen atom transfer, exhibits high chemo- and regioselectivity, acts catalytically, is mild towards the substrate and oxidized product, performs under strictly neutral conditions, and can be conveniently prepared from readily available commercial materials (Table 1).

Dimethyldioxirane is prepared from the reaction of acetone with caroate [Eq. (1)] under buffered conditions and subsequently isolated by distillation in the form

$$CH_3COCH_3 \quad + \quad KHSO_5 \xrightarrow[\text{pH} \sim 7.4,\, 5\text{ - }10\,°C]{H_2O,\ NaHCO_3} \quad \underset{CH_3}{\overset{CH_3}{>}}\!\!<\!\!\underset{O}{\overset{O}{\big|}}$$

$$(0.09\text{-}0.11\,\text{mol l}^{-1}) \tag{1}$$

---

* *Dedicated in admiration and friendship to Professor F. D. Greene III (Cambridge, Mass.) on the occasion of his 65th Birthday.*

Topics in Current Chemistry, Vol. 164
© Springer-Verlag Berlin Heidelberg 1993

Table 1. Comercial sources of potassium Mono-
peroxysulfate

$2 \cdot KHSO_5 \cdot KHSO_4 \cdot K_2SO_4$
(M.W. = 614)

Industrial Sources:
Degussa   Caroate
Dupont   Oxone
Interox   Curox
Peroxid-Chemie Curox

Price:
Industrial $ 4/kg
Aldrich $ 45/kg

---

of a $0.08-0.10 \text{ mol } l^{-1}$ acetone solution. A most convenient procedure is given below and the apparatus exhibit in Fig. 1.

Preparation of Acetone Solutions of Dimethyldioxirane [4]

A 4000-ml, three-necked, round-bottomed reaction flask was equipped with an efficient mechanical stirrer and a solid addition funnel, connected by means of a U-tube (I.D. 25 mm) to a two-necked receiving flask and cooled to $-78\,°C$ by means of a dry ice — ethanol bath (Fig. 1). The reaction flask was charged with a mixture of water (254 ml), acetone (192 ml), and $NaHCO_3$ (58 g) and cooled to $5-10\,°C$ with the help of an icewater bath. While vigorously stirring and cooling, solid caroate (120 g, 0.195 mol) was added in five portions at 3-min intervals. A moderate vacuum (80–100 Torr) was applied 3 min after the last addition, the cooling bath ($5-10\,°C$) removed from the reaction flask, and while vigorously stirring the dimethyldioxirane/acetone distillate (150 ml, $0.09-0.11 \text{ mol } l^{-1}$; ca. 4% yield) collected in the cooled ($-78\,°C$) receiving flask. The concentration of dimethyldioxirane was determined by oxidation of methyl phenyl sulfide to its sulfoxide,

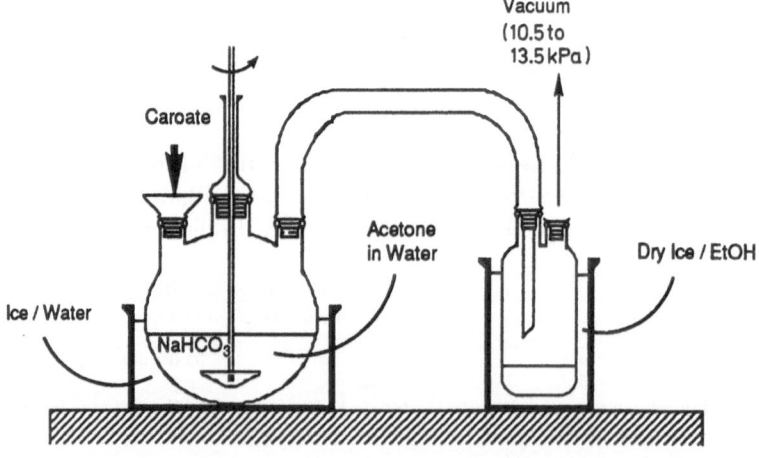

Fig. 1. Simplified apparatus for the preparation of isolated dimethyldioxirane

the latter quantitated by ^{1}H NMR. The acetone solution of dimethyldioxirane was dried over K_2CO_3 and stored in the freezer ($-20\,°C$) over molecular sieves (4 Å).

When the substrate and oxidized product tolerate hydrolytic conditions, the oxidation can be performed in situ. For this purpose we have found it advantageous to employ 2-butanone [5] instead of acetone as source of the dioxirane. Because of its partial solubility in water and excellent solvent properties no cosolvents such as CH_2Cl_2 or C_6H_6 are required. For convenience, in Table 2, we have summarized the reaction conditions and variables used for the dioxirane oxidations in the isolated and in situ modes.

Table 2. Reaction conditions and variables

	in situ	isolated
Dioxirane:	CH_2CH_3 CH_3 >C< O, O	CH_3 CH_3 >C< O, O
Scale:	up to moles	up to 100 mmoles
Stoichiometry:	◄——— equimolar to 10 fold ———►	
pH:	7.3–7.5	neutral
Solvent:	◄——— $R_2C=O$, CH_2Cl_2 ———►	
Temperature:	0 to $+30\,°C$	-80 to $+30\,°C$
Time:	hours	min to days

The general features of reactivity and selectivity of this novel oxidant are displayed in Table 3. It is significant that the fluoro derivative, i.e. methyl-(trifluormethyl)dioxirane [6], is at least 1000-fold more reactive than dimethyl-dioxirane. As a consequence, the fluorinated dioxirane oxidizes alkanes to the corresponding alcohols and/or ketones within minutes even at subambient temperatures [7].

As expected for an electrophilic oxidant such as dimethyldioxirane, the higher the degree of alkylation of the alkene, the faster the rate of epoxidation. Moreover, cis-alkenes react slightly faster than the *trans*-isomers [8]. Electron donors activate the double bond towards epoxidation, while electron acceptors deactivate it.

With regard to the stereochemistry of the oxygen transfer, the oxidation proceeds with preservation of the initial configuration, both for epoxidation and σ bond insertions (C−H, Si−H). The diastereoselectivity, as it is for peroxy acids, is rather low (a. 60:40). However, when sterically massive groups are present, as in the oxazoline (Table 3), diastereocontrol can be very high (>99:1). A typical case is the epoxidation of the norbornene in Table 3, which affords exclusively the *exo*-epoxide.

Table 3. Reactivity and selectivity of dioxiranes

REACTIVITY:

DIOXIRANES -

$$CF_3\text{-}C\overset{CH_3}{\underset{}{<}}\overset{O}{\underset{O}{|}} \quad \gg \quad CH_3\text{-}C\overset{CH_3}{\underset{}{<}}\overset{O}{\underset{O}{|}} \qquad \text{ca. 1000-fold}$$

OLEFINS -

DEGREE OF
SUBSTITUTION: $R_2C=CR_2 > R_2C=CHR > R_2C=CH_2 > RCH=CHR > RCH=CH_2$

cis - RCH=CHR > trans - RCH=CHR ;

TYPE OF
SUBSTITUTION: $R_3SiO- > RO- > (RO)_2\overset{O}{\overset{||}{P}}O- > R- > Ar- > RCO_2^- > X\text{-}\overset{O}{\overset{||}{C}}\text{-}$

$ED\text{-}\overset{|}{C}=\overset{|}{C}\text{-}ED > ED\text{-}\overset{|}{C}=\overset{|}{C}\text{-} > \underset{}{>}C=C\underset{}{<} > ED\text{-}\overset{|}{C}=\overset{|}{C}\text{-}EA > -\overset{|}{C}=\overset{|}{C}\text{-}EA$

CHEMOSELECTIVITY:

$$X(S,N,P) > C=C > H\text{-}\overset{|}{\underset{|}{C}}\text{-}OH \; ; \; H\text{-}\overset{|}{C}=O > H\text{-}\overset{|}{\underset{|}{C}}- > \text{Arenes} \; ; \; C\equiv C$$

STEREOCHEMISTRY:

Me $\overset{Ph}{\underset{Et}{C}}$ H → Me $\overset{Ph}{\underset{Et}{C}}$ OH   > 99 %

Me $\overset{Ph}{\underset{\alpha\text{-Naph}}{Si}}$ H → Me $\overset{Ph}{\underset{\alpha\text{-Naph}}{Si}}$ OH   > 99 %

DIASTEREOSELECTIVITY:

d.r.= 70:30   62:38   60:40   60:40   65:35

d. r. =   85:15   > 99:1   > 99:1

ENANTIOSELECTIVITY:

$$Me\overset{R^*}{\underset{}{C}}\overset{O}{\underset{O}{|}} \qquad R^* = Me\text{-}\overset{H}{\underset{Ph}{C}}- \qquad e.e. = 12 \%$$

To date, only a little work has been carried out on the enantioselective epoxidations by dioxiranes. For example, with 1-phenylethyl as asymmetric inductor about 12% e.e. has been reported [9].

2 Synthetic Applications

Scheme 1 shows the various types of oxyfunctionalizations which dimethyl-dioxirane is cabable of performing. These include epoxidations of π systems, oxidations of heteroatoms, and insertion into σ bonds. Here we will restrict

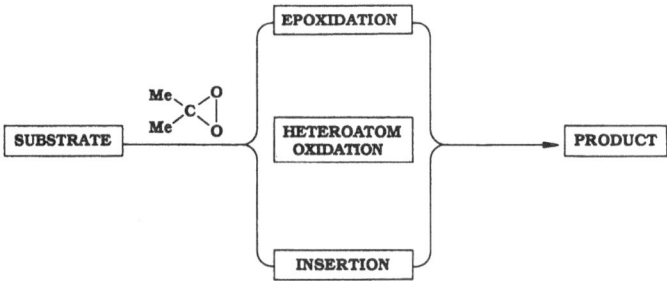

Scheme 1. Dimethyldioxirane oxyfunctionalization modes

ourselves to dimethyldioxirane oxidations performed in our laboratory during the last two years. Most of the work by others is to be found in existing reviews on the subject [3]. Concerning our own work, we present the results in the form of generic types, the details being given in the respective publications.

We classify the C=C π systems in terms of their reactivity as electron-rich substrates, i.e. those that bear electron donors (ED), and electron-poor substrates,

Formulas A

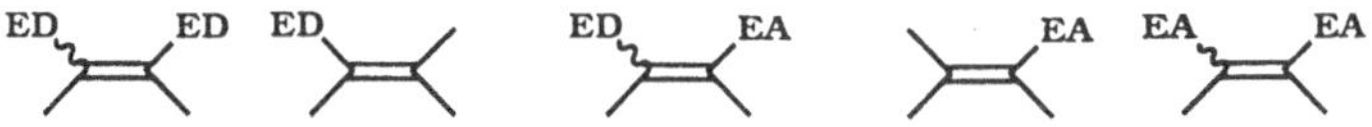

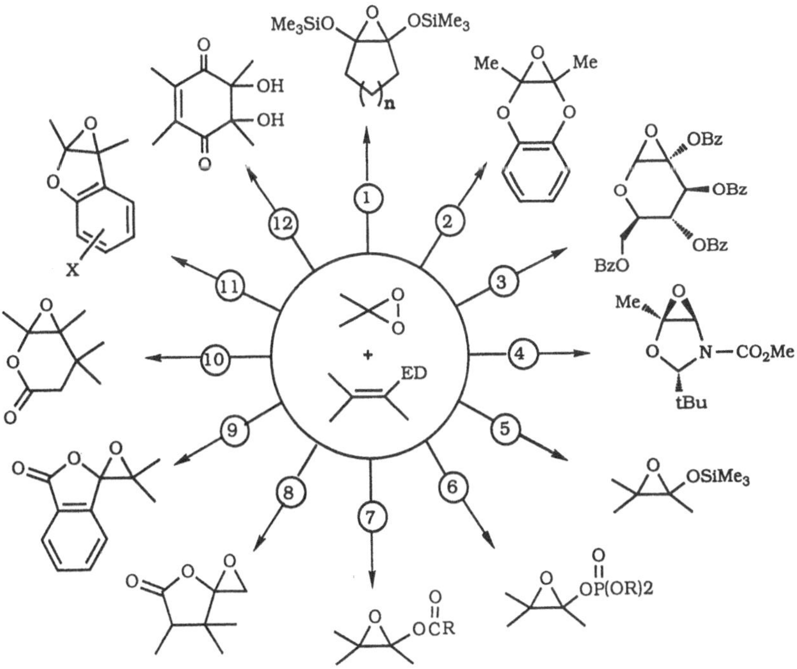

Scheme 2. Dimethyldioxirane oxidation of substrates with electron donors (ED)

49

Waldemar Adam and Lazaros Hadjiarapoglou

i.e. those that bear electron acceptors (EA). We shall briefly discuss these substrates in this reactivity order.

Scheme 2 summarizes the π systems with electron-donating substituents. These include 1,2-bis-trimethylsilyloxycycloalkenes [10] **1**; benzo-*p*-dioxin [10] **2**; tetrabenzyl-*O*-glycal [10] **3**; oxazoline [11] **4**; silyl enol ethers [12] **5**; enol phosphates [12b], [13] **6**; enol esters [12], [14] **7**; butyrolactones [14] **8**; phthalide [12] **9**, enol lactones [12] **10**; benzofurans [15] **11**; and hydroquinones [16] **12**. The labile epoxides were obtained in essentially quantitative yields and spectrally characterized.

Scheme 3 presents substrates which carry both electron-donating and electron-withdrawing substituents. These systems, in comparison with the previous set of substrates, were considerably less reactive so that longer reaction times and excess oxidant were necessary for complete epoxidations. In the rosette are displayed the oxidation of β-*oxo* enol phosphates [13] **13**, dihydrofuranone [17] **14**, β-alkoxycyclohexenones [17] **15**, alkoxymethylenecyclohexanones [17] **16**, benzalphthalide [12a] **17**, aurones [18] **18**, flavones [19] **19**, and isoflavones [18] **20**. Many of these epoxides, which have become available for the first time, constitute valuable building blocks for natural product chemistry.

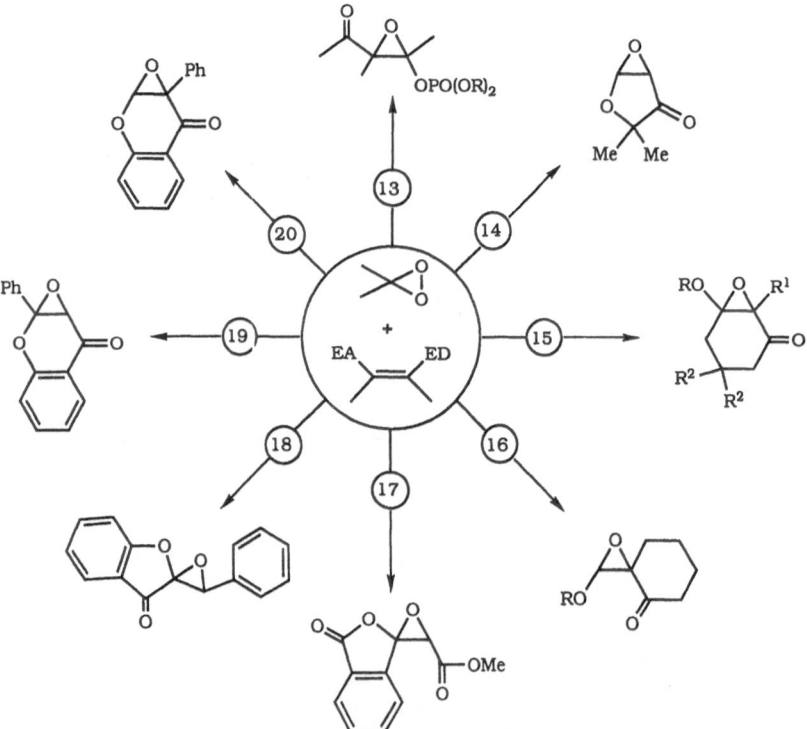

Scheme 3. Dimethyldioxirane oxidation of substrates with electron donors (ED) and electron acceptors (EA)

50

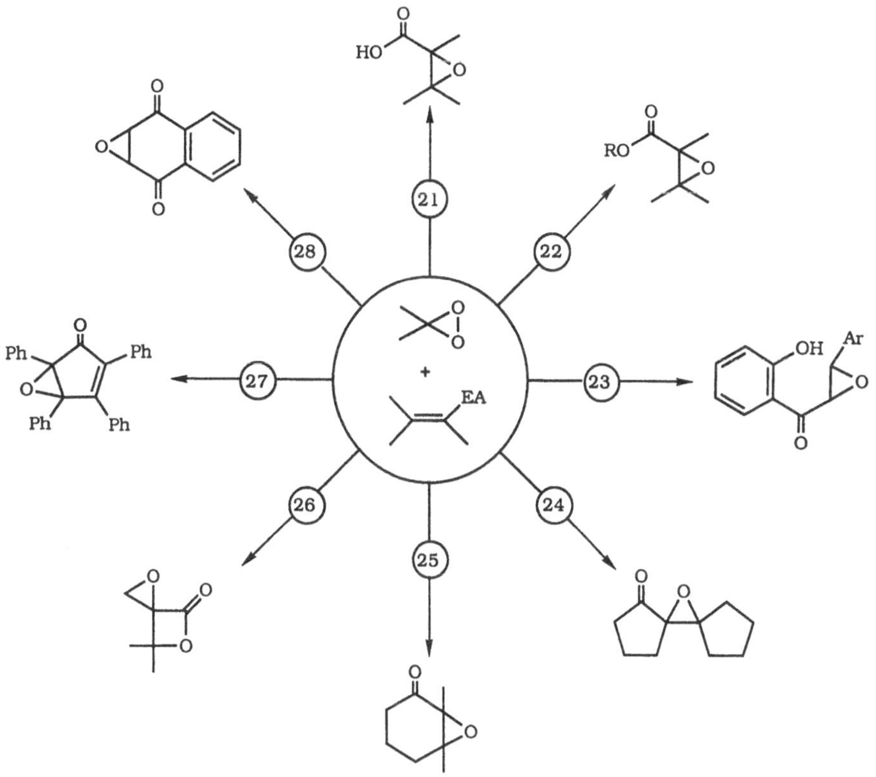

Scheme 4. Dimethyldioxirane oxidation of substrates with electron acceptors (EA)

Substrates with electron acceptors are given in Scheme 4. For these rather unreactive π-systems, besides longer reaction times and excess dioxirane, elevated temperature (ca. 25 °C) was essential for complete conversion into their corresponding epoxides. Fortunately, these epoxides, which were all obtained quantitatively, were sufficiently stable for rigorous characterization. The specific substrate types include α,β-unsaturated acids [20] **21** and esters [20] **22**; hydroxychalcones [21] **23**; α,β-enones [20] **24**; 2-cyclohexenones [20] **25**; methylene-β-lactones [22] **26**; tetracyclone [20] **27**; and naphthaquinone [16] **28**. The particular advantage of the dioxirane methodology is that labile functional groups, e.g. the phenolic moiety in the chalcones, do not require protection.

Examples of substrates [23] with heteroatom double bonds (C=X) are presented in the rosette of Scheme 5. Here we feature the efficient denitrogenation of diazoalkanes, diazoquinones, the hydrazone of benzil, and benzophenone oxime into the respective carbonyl products [23]. On the other hand, the bis-oxime afforded furoxane, while the phosphorane led to methyl maleate on elimination of triphenylphosphine oxide [23]. Thiones are oxidized to the corresponding S-oxides and the iodonium ylide yields the labile *vic*-trione [23].

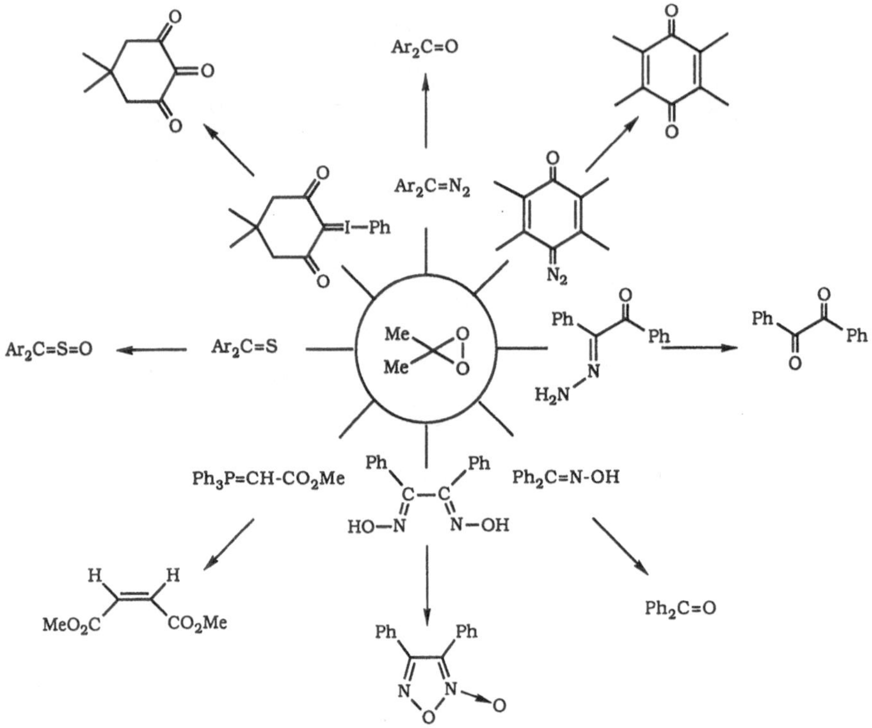

Scheme 5. Dimethyldioxirane oxidation of substrates with $C=X$ functionalities

Oxygen transfer to sulfur, which leads to sulfoxides and/or sulfones are displayed in Scheme 6. Excess dioxirane converts bis-thiophenylmethane into its bis-sulfone [23] **29** quantitatively. When using stoichiometric amounts of dioxirane, thioenol ethers are transformed into their sulfoxides [23] **30**, but excess dioxirane leads to the sulfone [23] **31**. As expected, the thioisoflavone [24] **32**, the allene sulfoxide [25] **33**, and the bis-methylthiobutadiene [26] **34** gave the corresponding sulfones even when excess dioxirane was employed. Of potential synthetic value is the direct oxidation of thioesters to oxo sulfones [27] **35**. Finally, diphenyl disulfide afforded the S-thiosulfinate [23] **36** as major product when dioxirane was used in stoichiometric quantity.

The unusual reactivity of dioxiranes is impressively exhibited in their ability to insert into $C-H$ bonds (Scheme 7) [28]. Thus, tertiary alkanes are oxidized to their respective alcohols [29]. In the example shown, the insertion took place with complete retention of configuration at the chirality center. 1,3-Dicarbonyl derivatives [30] are hydroxylated with high efficiency, but more than likely the intermediary enol is being oxyfunctionalized. Secondary alcohols are transformed into ketones, a specific example is the oxidation of the epoxy alcohol in the rosette [31]. In an attempt to epoxidize the hydroxy acrylic ester [22], the epoxy 1,3-dicarbonyl product was obtained, although in low yield in accord with its rather reluctant nature towards oxidation.

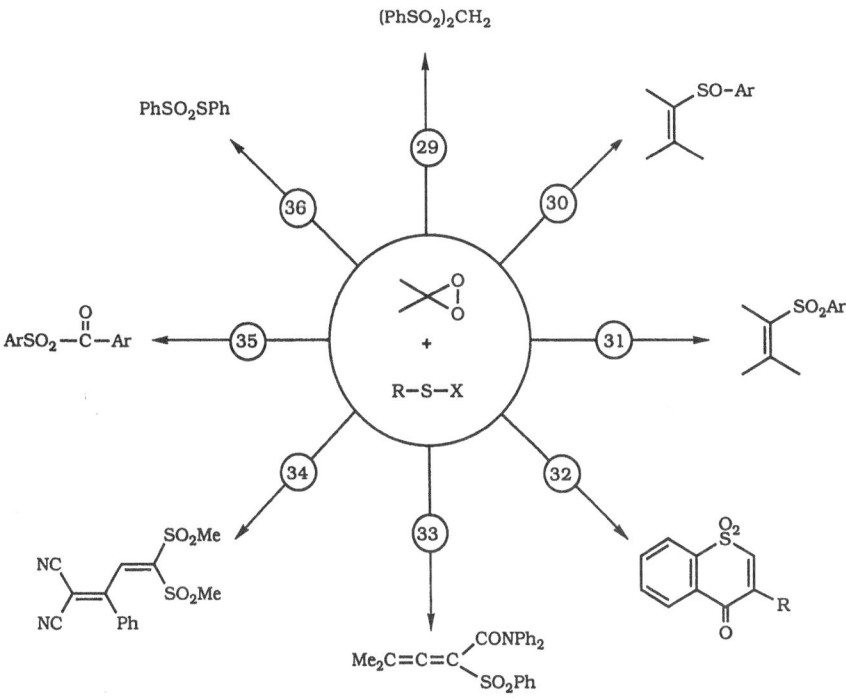

Scheme 6. Dimethyldioxirane oxidation of substrates containing sulfur functionalities

While ethers react only slowly with dimethyldioxirane, they are efficiently hydroxylated by methyl(trifluoromethyl)dioxirane even at low temperatures. Thus *t*-butyl methyl ether [32] was converted to *t*-butyl alcohol through its hemiacetal. On the other hand, tetrahydrofuran gave butyrolactone [32] in which presumably the intermediary cyclic hemiacetal was oxidized to the lactone by an additional C−H insertion. The ketal was degraded into 2-butanone and the orthoformate into diethyl carbonate [32]. The latter transformation may serve useful for deketalation under neutral conditions.

During the last five years the oxidation of organic compounds by dioxiranes has been extensively investigated, but relatively little work has been reported on organometallic substrates. To fill this gap, we have ventured into this challenging and promising area. Our preliminary results in Scheme 8 convincingly illustrate the advantages of dimethyldioxirane for selective epoxidations, heteroatom oxidations, and σ bond insertions. Thus, the silyl enol ethers [12] afforded the corresponding epoxides, while the enolates of lithium, sodium and titanium gave, in high yield, the α-hydroxy carbonyl products [30]. Here it is significant to mention that for optimal results, the dioxirane must be administered by inverse addition at −78 °C to the highly reactive enolates of the alkali metals.

Allyl stannanes led to rather labile epoxides on treatment with dimethyldioxirane, which on warming to room temperature generated the corresponding

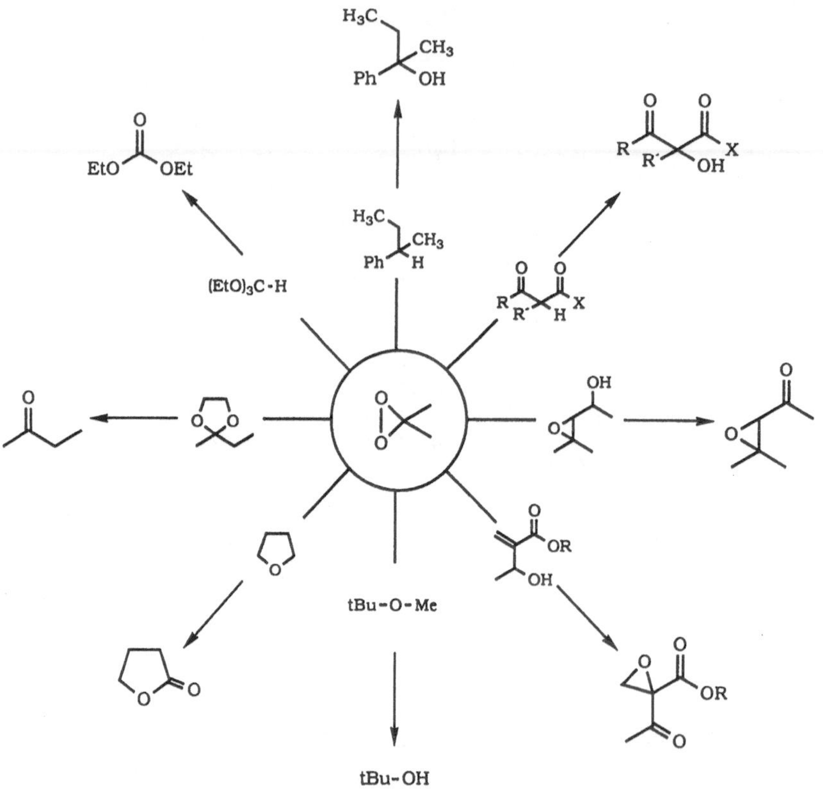

Scheme 7. Dimethyldioxirane C−H insertions

allyl alcohol by elimination of trialkyltin hydroxide [33]. The *N,N*-bis(trimethyl-silyl)enamines gave on inverse addition of dioxirane at −78 °C the first obervable enamine epoxides, which decomposed at elevated temperature [34].

In view of the current interest in iron complexes for organic synthesis, we have employed dimethyldioxirane for epoxidation purposes. For example, the quite reluctant iron tricarbonyl diene complex with the alkenyl side chain led to the desired epoxide in moderate yield [35]. Also the isoprenyl-substituted ferrocene could be oxyfunctionalized without any difficulties [36].

Examples of sulfur oxidation constitute the sulfone of the ruthenium complex and the sulfoxide of the tungsten derivative [37]. In the latter, an excess of dioxirane effected the oxidation to the corresponding sulfone. Oxyfunctionalizations of this type are unprecedented.

Finally, dimethyldioxirane is a convenient oxidant for converting silanes into their silanols by Si−H insertion. When optically active trialkylsilanes are used, the Si−H insertion proceed with complete retention of configuration [38]. The advantage of this highly selective method of hydroxylating silanes is the fact that under the strictly neutral conditions no siloxanes are produced. Likewise,

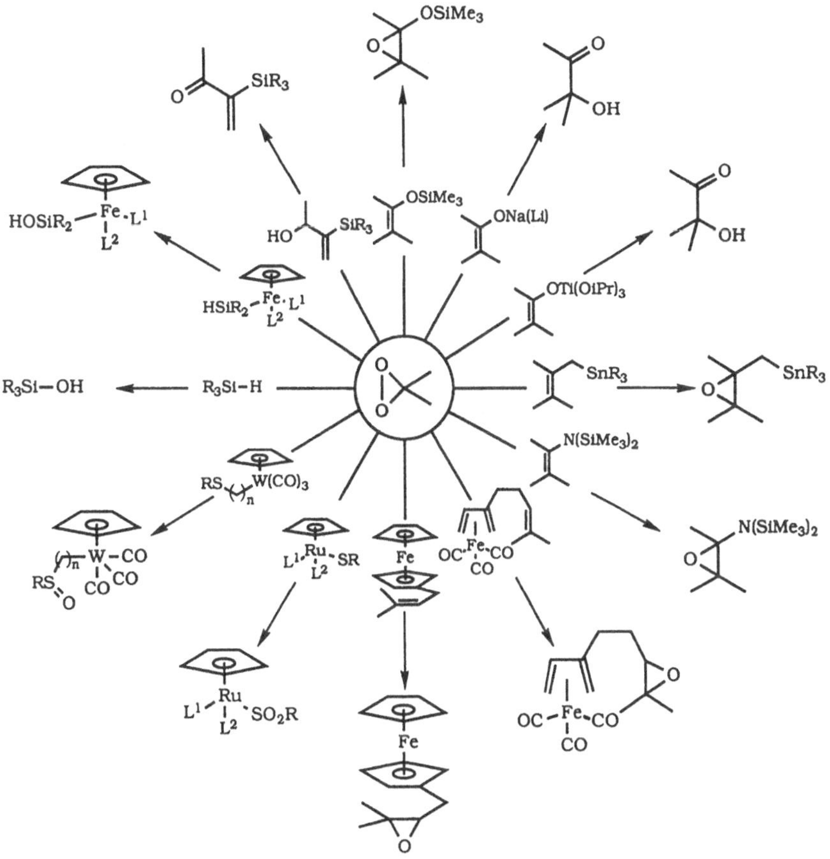

Scheme 8. Dimethyldioxirane oxidations of organometallic substrates

iron-substituted silanes afforded, under these mild conditions, the corresponding silanols in high yield [39]. Surprising was the hydroxy vinyl silane, which underwent C−H insertion rather than epoxidation [40].

3 Mechanistic Considerations

The previous section amply demonstrates the advantages of dimethyldioxirane as an oxygen transfer agent for preparative oxidation chemistry compared to other oxidants such as peroxy acids and alkaline hydrogen peroxide; the synthetic applications are literally endless. Nonetheless, the mechanistic details of the oxygen transfer are to date still inadequately understood. Presently available experimental data [3] such as sterochemistry, kinetics, activation parameters, isotope effects, reactivity patterns, etc. are consistent with the complex "butterfly" transition state **A**, initially proposed for peroxy acids as oxygen atom donors, and the novel diradical-like transition state **B** (Scheme 9).

Waldemar Adam and Lazaros Hadjiarapoglou

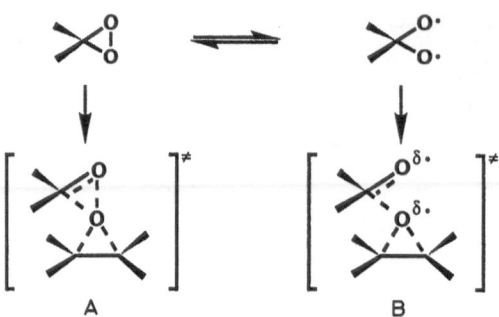

Scheme 9. Mechanism of oxygen transfer by intact dioxirane (A) and dioxyl diradical (B)

In **A** the intact dioxirane delivers the oxygen atom to the alkene, while in **B** it is the ring-opened dioxirane, i.e. the 1,3-dioxyl diradical. All other geometrical features are essentially the same. The latter hypothesis demands facile homolysis of the peroxide bond under the oxygen transfer conditions. A theoretical estimate [41] of the reaction enthalpy for ring-opening of the dioxirane into its 1,3-diradical is ca. 10 kcal/mol, so that the activation energy for this process is expected to be low, certainly less than 15 kcal/mol. In fact, recent computations [42] reveal that the direct oxygen atom transfer by the intact dioxirane, i.e. the "butterfly" transition state **A**, requires quite high activation energy ($E_a > 15$ kcal) and imply that the $O-O$ bond is essentially broken. Whether this derives from alkene-induced homolysis or spontaneous cleavage of the peroxide bond to generate the dioxyl diradical first cannot be answered at this time because the energetics for the trajectory of the oxygen transfer by the dioxyl diradical to the alkene must be assessed. Apparently this is still a difficult task for such an open-shell species [43]. In this context it is of interest to mention that calculations [44] on the related dioxetanes suggest that the activation energy for peroxide bond homolysis into the corresponding 1,4-dioxyl diradical [Eq. (2)] is negligible ($E_a \sim 0$ kcal/mol).

$$R\text{---}\underset{R\ R}{\overset{O-O}{||}}\text{---}R \longrightarrow R\text{---}\underset{R\ R}{\overset{\overset{\bullet}{O}\ \overset{\bullet}{O}}{||}}\text{---}R \tag{2}$$

The difference in the reaction profiles that involve transition states **A** and **B** for the oxygen transfer are quite subtle, i.e. merely the $O-C-O$ bond angle in the attacking oxygen atom donor. In the intact dioxirane this bond angle is ca. 60°, while in the dioxyl diradical it is at most ca. 110°.

A reaction path which is not consistent with the experimental facts, is the addition of the dioxirane or its dioxyl diradical to produce the 1,5-diradical **C** [Eq. (3)]. Such a diradical would be expected to be sufficiently long-lived to rotate

$$\tag{3}$$

C

56

about the carbon radical site and afford detectable amounts of stereo-randomized epoxides. However, all epoxidations by dioxiranes known to date proceed with rigorous preservation of the initial alkene configuration. Moreover, some 1,3-dioxolane would be expected as a product, but so far no such cycloaddition has been observed.

Irrespective of these interesting queries on the mechanistic details of the oxygen transfer by dioxiranes, the even more puzzling question must be raised as to why the readily accessible dioxyl diradical does not rearrange into the corresponding ester [Eq. (4)], especially if it is realized that this reaction channel is exothermic

$$
\underset{R}{\overset{R_{_{\prime\prime}}}{\diagdown}}\!\!\!\underset{O}{\overset{O}{\diagup}} \quad\longrightarrow\quad \underset{R}{\overset{R_{_{\prime\prime}}}{\diagdown}}\!\!\!\underset{O\bullet}{\overset{O\bullet}{\diagup}} \quad\longrightarrow\quad \underset{R}{\overset{O}{\diagdown}}\!\!\!\underset{O}{\overset{\parallel}{\diagup}}\!\!R \tag{4}
$$

by about 120 kcal/mol [45]. The experimental facts are that thermally little if any ester product is formed on heating dioxiranes in solution [46]. On the other hand, the photolysis under matrix isolation constitutes early experimental evidence [47] for the rearrangement of dioxiranes into esters. The fact that in a recent study [48] matrix-phase photolysis of the methyl(trifluoromethyl)dioxirane afforded not only methyl trifluoroacetate, but also carbon dioxide and 1,1,1-trifluoroethane [Eq. (5)], provides cogent experimental data that the corresponding dioxyl dira-

$$
\underset{CH_3}{\overset{CF_3}{\diagdown}}\!\!\!\underset{O}{\overset{O}{\diagup}} \quad\overset{h\nu}{\longrightarrow}\quad \underset{CH_3}{\overset{CF_3}{\diagdown}}\!\!\!\underset{O\bullet}{\overset{O\bullet}{\diagup}} \quad\longrightarrow\quad [CF_3CO_2^\bullet \cdot CH_3]
$$

$$
CF_3CO_2CH_3 \qquad CF_3\text{-}CH_3 + CO_2 \tag{5}
$$

dicals intervene. How else could the radical coupling product CF_3-CH_3 be formed? Alkoxy radical-type β-scission of the dioxyl species generates the carboxy-methyl radical pair, which on one hand couples to give the ester product and on the other hand decarboxylates to afford CO_2 and CF_3CH_3.

Electronic reasons appear to be responsible for dioxiranes not being rearranged into ester products on heating, despite the fact that such a rearrangement constitutes a rather exothermic process [45]. To understand the thermal persistence of these most strained cyclic peroxides, we must take a closer look at the electronic configurations of these diradicals.

A theoretical study [41] considered the three possible electronic states [σ, σ], [σ, π] and [π, π] for these dioxyl species (Scheme 10). These GVB (3)-CI calculations predicted the stability order [π, π] > [σ, π] > [σ, σ]. Examination of the orbital orientation in these three configurations reveals that in the [σ, σ]-type 1,3-diradical, the radical lobes and the σ bond of the migrating alkyl group are essentially orthogonal. Such a configuration should not be prone to rearrange into the ester by a 1,2-alkyl shift; instead, one expects it to reclose to the dioxirane. Therefore, on thermal activation, ring-opening of the peroxide bond affords the [σ, π]

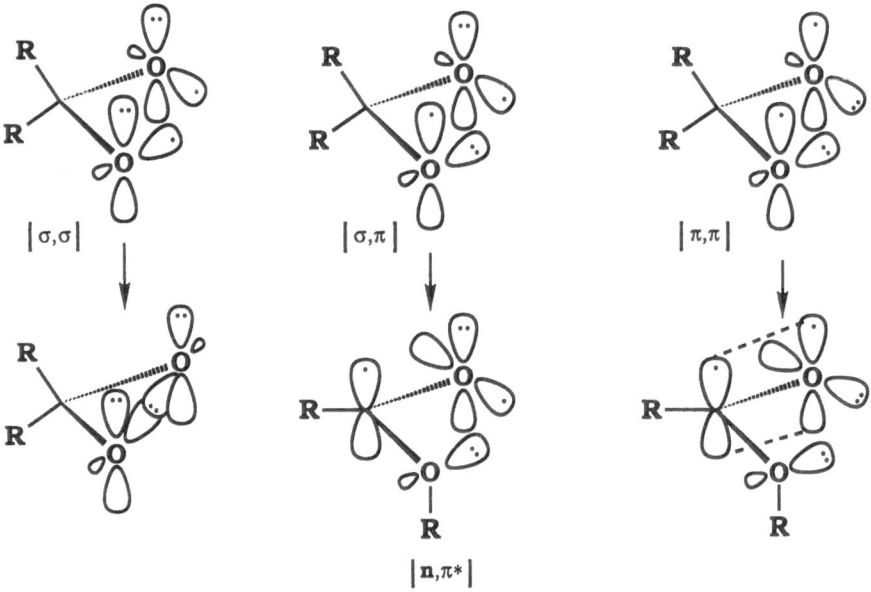

Scheme 10. Qualitative orbital representation for the rearrangement of the [σ, σ], [σ, π] and [π, π] electronic configurations of the dioxyl diradicals

dioxyl diradical, but since rearrangement to the ester requires further thermal activation, the preferred course of action is recyclization into the dioxirane (Fig. 2).

This ring-opening and reclosure process does not apply for the [σ, π] diradical configuration, because at least one of the radical lobes is coplanar with the σ

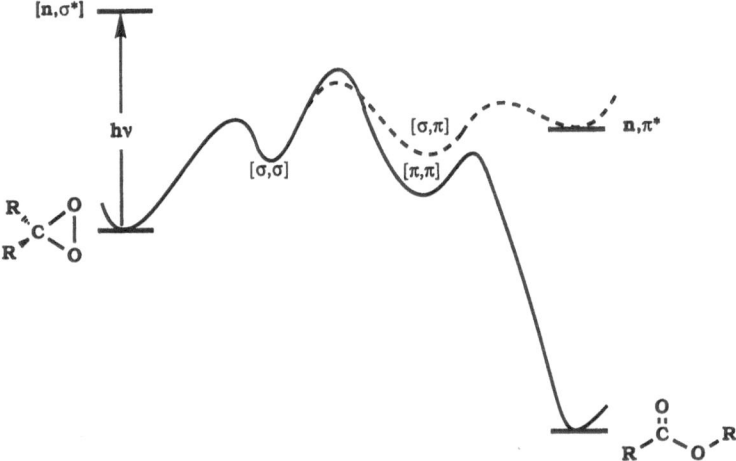

Fig. 2. Reaction profiles for the transformation of the [σ, σ], [σ, π] and [π, π] electronic configurations of the dioxyl diradicals

bond of the migrating alkyl group (Scheme 10). However, consideration of the non-crossing rule [49] suggests that the [σ, π] configuration will lead to the n, π* excited ester product on alkyl migration (Fig. 3). On the basis of thermochemical estimates, we suggested previously [3b] that such a transformation is feasible, but competing lower energy pathways appear to be more probable, e.g. β scission to generate a carboxy-alkyl radical pair. This is exactly what is observed in the long-wavelength photolysis of dioxiranes under matrix isolation [Eq. (5)]. Indeed, the lowest excitation of the dioxirane is of the n_, σ* type (the antibonding combination n_ of the lone pairs perpendicular to the molecular plane have π* type symmetry [50], as shown in Fig. 4, which correlates with the [σ, π] diradical). This is graphically presented in the correlation diagram of Fig. 3. Since n_, σ* excitation of the dioxirane predicts generation of the n, π* ester product through rearrangement of the intervening [σ, π] diradical, it should be of interest to confirm this through chemiluminescence studies under matrix isolation. Nevertheless, a prominent pathway [Eq. (5)] is β-scission of the [σ, π] diradical into a carboxy-alkyl radical pair, which is the source of some of the ester product through in-cage radical coupling.

A favorable situation for the rearrangement of dioxiranes into esters obtains for the [π, π] diradical. For this electronic configuration the radical lobes are perfectly aligned for rearrangement with the σ bond of the migrating alkyl group (Scheme 10). In view of the high exothermicity for this process (Fig. 3), the rearrangement must be quite efficient. It is, therefore, not surprising that in the gas phase pyrolysis at 500 °C methyl(trifluoromethyl)dioxirane afforded exclusive-

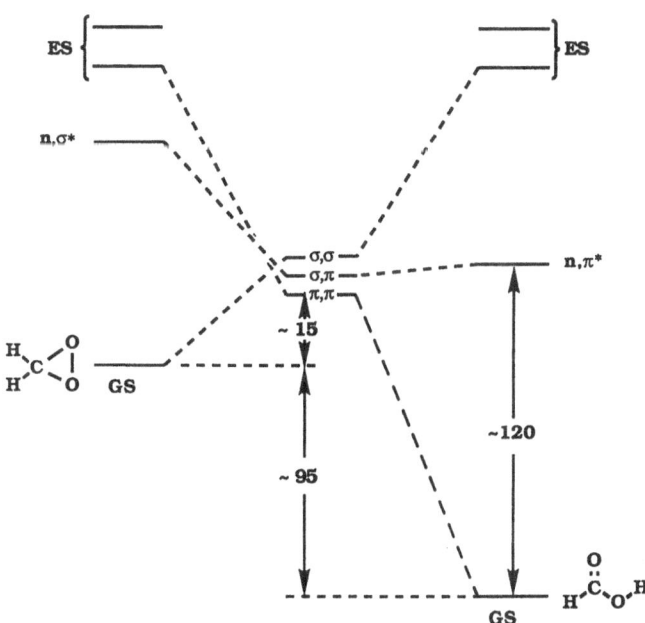

Fig. 3. Correlation diagram for the rearrangement of dioxirane

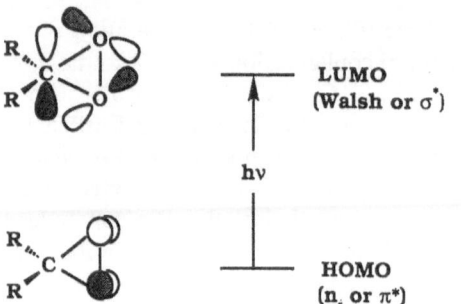

LUMO
(Walsh or σ˙)

hν

HOMO
(n₋ or π*)

Fig. 4. Lowest excitation (n₋, σ*) for dioxiranes

ly methyl trifluoroacetate [46]. At these elevated temperatures, the initially formed [σ, σ] diradical is transformed into its [π, π] electronic isomer and the latter rearranges into the ester by methyl migration. That a methyl 1,2-shift is preferred over a trifluoromethyl one is a consequence of the significantly lower (by ca. 10 kcal/mol bond) energy for the former.

Scheme 11 summarizes the various unimolecular transformations of the [σ, σ]-, [σ, π]-, and [π, π]-type dioxyl diradicals, which have been disclosed by our qualitative analysis of the orbital orientations in these electronic isomers. The answer to the original question as to the reasons for the persistence of dioxiranes rests on the appreciable electronic barrier towards rearrangement into the ester for the thermally produced [σ, σ] diradical; instead, the [σ, σ] dioxyl species recloses to the dioxirane! Consequently, these most highly strained cyclic peroxides can be prepared, isolated, handled, and utilized for synthetic purposes even under ambient conditions. On photochemical activation (n₋, σ* excitation), however, the [σ, π] diradical results, which prefers β-scission into a carboxy-alkyl radical pair rather than rearranging into the ester product. For the latter process,

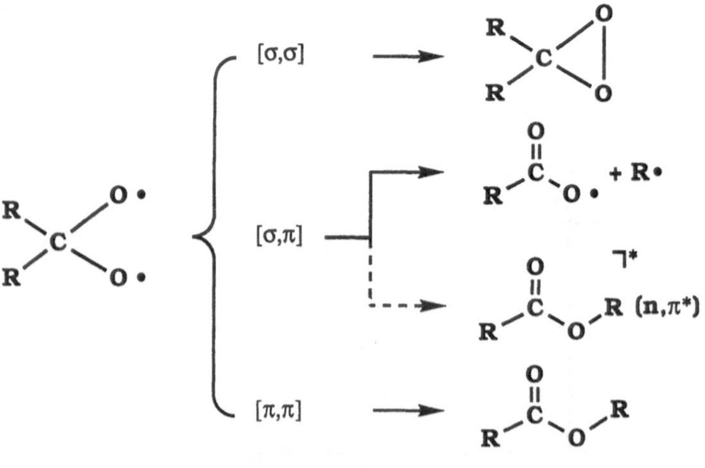

Scheme 11. Transformations of the [σ, σ], [σ, π] and [π, π] configurations of the dioxyl diradicals

conservation of spatial symmetry (a problem of $\sigma-\pi$ crossing) would necessitate the formation of the n, π^*-excited ester product, which is presumably of higher energy than β-scission. Finally, the $[\pi, \pi]$ configuration of the dioxyl diradical, accessible under pyrolysis conditions at elevated temperatures, is, in view of the optimal alignment of the involved orbitals, destined for rearrangement into the ester by an alkyl 1,2 shift. We postulate that the dioxyl diradical may serve as a viable oxygen atom donor in oxygen transfer reactions such as epoxidation, hetero-atom oxidation and σ bond insertion. Careful theoretical work will have to elucidate whether the intact dioxirane and/or its ring-opened dioxyl diradical are the oxidants in the observed oxyfunctionalizations.

Acknowledgements The generous financial support by the DFG (SFB 347) "Selektive Reaktionen Metall-aktivierter Moleküle" and the Fonds der Chemischen Industrie is gratefully appreciated.

4 References

1. Recipient of the 1991 Interox Junior Awart; part of this material was presented by Lazaros Hadjiarapoglou as the Award Address at the 3. ORPEC Conference, München, April 29, 1991
2. Murray RW, Jeyaraman R (1985) J Org Chem 50: 2847
3. a. Murray RW (1988) In: Liebmann JF, Greenberg A (eds), Molecular structure and energetics. Unconventional chemical bonding, vol 6, VCH Publishers, New York, p 311
   b. Adam W, Curci R, Edwards JO (1989) Acc Chem Res 22: 205
   c. Murray RW (1989) Chem Rev 89: 1187
   d. Curci R (1990) In: Baumstark AL (ed) Advances in oxygenated processes, vol 2, JAI Press, Greenwich, p 1
   e. Adam W, Curci R, Hadjiarapoglou L, Mello R In: Organic Peroxides, Ando W (ed), John Wiley-Interscience (in press)
4. Adam W, Bialas J, Hadjiarapoglou L (1991) Chem Ber 124: 2361 (in press)
5. a. Adam W, Hadjiarapoglou L, Smerz A (1991) Chem Ber 124: 227
   b. Adam W, Hadjiarapoglou L, Siegmeyer R German Patent Application p 40 23 741.9
6. Mello R, Fiorentino M, Sciacovelli O, Curci R (1988) J Org Chem 53: 3890
7. Mello R, Fiorentino M, Fusco C, Curci R (1989) J Am Chem Soc 111: 6749
8. Baumstark AL, Vasquez PC (1988) J Org Chem 53: 3437
9. Curci R, Fiorentino M, Serio MR J Chem Soc Chem Commun 1984: 155
10. Adam W, Hadjiarapoglou L, Wang X (1991) Tetrahedron Lett 32: 1295
11. Adam W, Hadjiarapoglou L, Seebach D (unpublished results)
12. a. Adam W, Hadjiarapoglou L, Wang X (1989) Tetrahedron Lett 30: 6497
    b. Adam W, Hadjiarapoglou L, Jäger V, Klicic J, Seidel B, Wang X (1991) Chem Ber 124: 2377
13. Adam W, Hadjiarapoglou L, Klicic J (1990) Tetrahedron Lett 31: 6517
14. Adam W, Hadjiarapoglou L, Jäger V, Seidel B (1989) Tetrahedron Lett 30: 4223
15. a. Adam W, Hadjiarapoglou L, Mosandl T, Saha-Möller C, Wild D (1991) Angew Chem 103: 187
    b. Adam W, Hadjiarapoglou L, Mosandl T, Saha-Möller C, Wild D (1991) J Am Chem Soc 113: 8005
    c. Sauter M Diplomarbeit (1991) University of Würzburg
    d. Adam W, Bialas J, Hadjiarapoglou L, Sauter M (1992) Chem Ber 124: in press
16. Adam W, Schönberger A (1992) Tetrahedron Lett 33: 53

17. Adam W, Hadjiarapoglou L (1990) Chem Ber 123: 2077
18. Adam W, Hadjiarapoglou L, Levai A (1992) Synthesis in press
19. a. Adam W, Golsch D, Hadjiarapoglou L, Patonay T (1991) Tetrahedron Lett 32: 1041
    b. Adam W, Golsch D, Hadjiarapoglou L, Patonay T (1992) J Org Chem 57: in press
20. Adam W, Hadjiarapoglou L, Nestler B (1990) Tetrahedron Lett 31: 331
21. Adam W, Bialas J, Hadjiarapoglou L, Patonay T Synthesis 1992: in press
22. Prechtl F Diplomarbeit 1989 University of Würzburg
23. Adam W, Hadjiarapoglou L (unpublished results)
24. Adam W, Hadjiarapoglou L, Levai A (unpublished results)
25. Adam W, Hadjiarapoglou L, Saalfrank R (unpublished results)
26. Adam W, Hadjiarapoglou L, Peseke K (unpublished results)
27. Adam W, Hadjiarapoglou L (1992) Tetrahedron Lett 33: in press
28. Murray RW, Jeyaraman R, Mohan L (1986) J Am Chem Soc 108: 2470
29. Adam W, Curci R, Gonzalez Nuñez ME, Mello R (1992) J Org Chem 57: in press
30. Adam W, Prechtl F (1991) Chem Ber 124: 2369
31. Adam W, Abou-Elzahab M, Saha-Möller CR Liebigs Ann Chem 1991: 445
32. Adam W, Curci R, Gonzalez Nuñez ME, Mello R (unpublished results)
33. Klug P, Diplomarbeit 1991 University of Würzburg
34. Adam W, Reißig H-U, Voerckel V (unpublished results)
35. Adam W, Azzena U (unpublished results)
36. Schuhmann R, Diplomarbeit 1992 University of Würzburg
37. Adam W, Frisch J, Prechtl F, Schenk WA Inorg Chem submitted
38. Adam W, Curci R, Mello R (1990) Angew Chem 102: 916
39. Adam W, Azzena U, Hindahl K, Malisch W, Prechtl F Chem Ber (submitted)
40. Adam W, Richter M (unpublished results)
41. Harding LB, Goddard III WA (1978) J Am Chem Soc 100: 7180
42. Bach RD (private communication)
43. Cremer D (private communication)
44. Reguero M, Bernardi F, Bottoni A, Olivucci M, Robb MA (1991) J Am Chem Soc 113: 1566
45. Adam W, Curci R (1981) Chim Ind (Milan) 97: 3004
46. Adam W, Curci R, Gonzalez Nuñez ME, Mello R (1991) J Am Chem Soc 113: 7654
47. Sander W (1990) Angew Chem 102: 362
48. Adam W, Bucher G, Hadjiarapoglou L, Sander W (unpublished results)
49. Wadt WR, Goddard III WA (1975) J Am Chem Soc 97: 3004
50. Cremer D (1983) In: Patai S (ed) The chemistry of peroxides, Wiley-Interscience, New York, p 1

Enantioselective Epoxidation with Peroxidic Oxygen

Eugen Höft

Central Institute of Organic Chemistry, Rudower Chaussee 5, O-1199 Berlin-Adlershof, FRG

Dedicated to Professor Alfred Rieche on the occasion of his 90th birthday

Table of Contents

Optically active epoxides are versatile and important intermediates for the preparation
of homochiral oxygen-containing compounds. The main route to such optically active
epoxides is the enantioselective epoxidation of olefins by reagents containing peroxidic
oxygen. Besides the application of optically active oxidizing agents, hydrogen peroxide in
the presence of optical auxiliaries, and molybdenum peroxo complexes with chiral ligands,
the Sharpless epoxidation of allylic alcohols using *t*-butyl hydroperoxide in the presence of
titanium isopropoxide and L-(+)- or D-(−)-dialkyl tartrate ester is the most practicable
method for the preparation of homochiral epoxy alcohols. This development is reviewed
briefly followed by own results concerning the Sharpless epoxidation of C_5-allylic alcohols
and the use of various hydroperoxides in this reaction. It is known that racemic allylic
alcohols can be kinetically resolved by the asymmetric epoxidation procedure. Now
preliminary results can be presented regarding the kinetic resolution of racemic hydroperoxides via Sharpless epoxidation and a new method for determining of the ratio of
enantiomers in chiral hydroperoxides by HPLC is described.

Topics in Current Chemistry, Vol. 164
© Springer-Verlag Berlin Heidelberg 1993

1 Introduction

The selectivity of a chemical reaction is a very important criterion. Besides the chemo- and regioselectivity, the stereoselectivity, i.e. the favored or excluded formation of one or several stereoisomers in the course of a chemical reaction, plays an important role. If there is a formation of (S)- and (R)-enantiomers from a prochiral compound, an enantioselective reaction takes place. What are the reasons for the growing interest in enantioselective reactions and preparation of homochiral compounds? Firstly, it is certainly the wish of the chemist to imitate the ability of nature by stereospecific synthesis in the laboratory. Secondly, there are some practical and economic reasons: many natural products and a great number of synthetic drugs have a chiral structure and the enantiomers can differ markedly in their biological activity. Sometimes only one of the enantiomers exhibits the wanted optimal activity, while the other is less active or totally inactive, or even toxic.

Therefore, especially in the case of drugs having a chiral centre, it has been necessary to prepare both antipodes to determine their biological activity and to use only that enantiomer which possesses the desired and optimal activity.

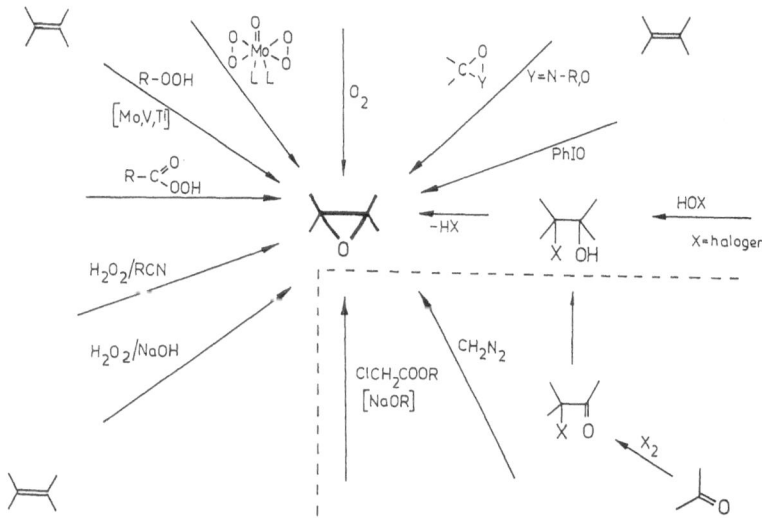

Scheme 1

Important intermediates for the synthesis of such biologically active compounds are epoxides. In Scheme 1, the well-known methods for the preparation of epoxides are shown. In view of the topic of this paper those methods are of special interest where the oxygen transfer to olefins takes place by a reagent with peroxidic oxygen.

Applying these methods for the epoxidation of prochiral olefins without additional measures racemic epoxides are obtained. In most cases, the idea is to make the method stereoselective and thus obtain pure or enriched enantiomers of epoxides by using chiral reagents or by addition of optically active auxiliaries. Some of the results obtained by various groups will be discussed.

2 Enantioselective Epoxidation of Alkenes

2.1 Use of Chiral Peroxy Acids and Hydrogen Peroxide in the Presence of Chiral Catalysts

Chiral peroxy acids, such as monoperoxycamphoric acid, have been used in a classical Prilezhaev reaction for the preparation of chiral epoxides [1], however, the enantiomeric excess (e.e.) of the epoxides formed was below 10%.

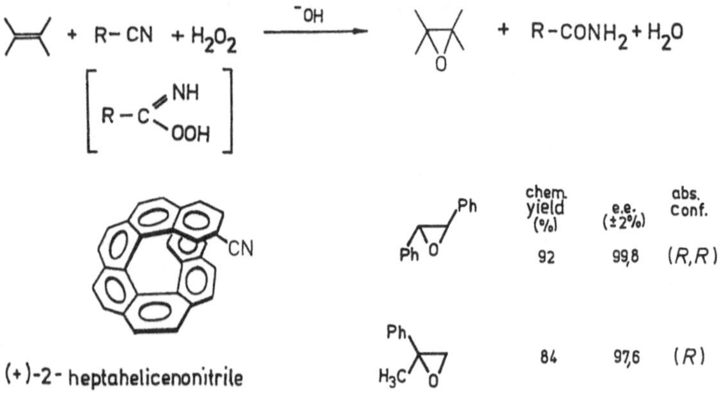

Scheme 2

Hydrogen peroxide can be used directly as a reagent for epoxidation. In the presence of nitriles, an iminoperoxy acid is formed in solution which can be successfully employed as an epoxidizing agent at pH 7.5–8 [2]. Normally acetonitrile, trichloroacetonitrile or benzonitrile are used as nitriles. Starting from an optically active heptahelicenonitrile in the epoxidation of (E)-stilbene and α-methylstyrene a surprisingly high e.e. was obtained with 98% hydrogen peroxide at room temperature [3].

Electrondeficient olefins can be epoxidized by hydrogen peroxide in the presence of bases. This method, known as the Weitz-Scheffer reaction, is used especially for the epoxidation of α,β-unsaturated ketones. Wynberg et al. [4] epoxidized a series of such ketones with hydrogen peroxide in the presence of quaternary ammonium salts of optically active alkaloids. Especially when phase-transfer conditions were used they yielded an e.e. of about 50% (Scheme 3).

In connection with investigations in the vitamin K series and considering the fact that some quinones and quinone epoxides are of importance in metabolic processes, the preparation of chiral quinone epoxides is of great interest. Such compounds were obtained with an e.e. of 45% by epoxidation of quinones using hydrogen peroxide in an alkaline medium in the presence of quininium benzyl chloride (QBC) [5].

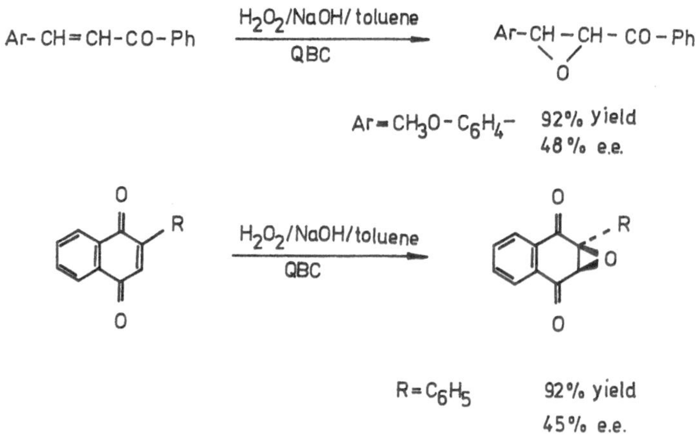
Ar-CH=CH-CO-Ph
H2O2/NaOH/toluene
QBC
Ar-CH-CH-CO-Ph
O
Ar=CH3O-C6H4-
92% yield
48 % e.e.
O
R
H2O2/NaOH/toluene
QBC
O
R
O
O
R=C6H5
92% yield
45% e.e.

Scheme 3

Juliá et al. [6] used readily available synthetic chiral polypeptides, such as poly (*S*)-alanine, as optical auxiliaries instead of salts of alkaloids (Scheme 4). So they were able to achieve the epoxidation of chalkones in high optical yields. The catalytically active polypeptide was synthesized according to Scheme 4. The use of simple polypeptides as "synthetic enzymes" may be a valuable tool for performing stereoselective epoxidations.

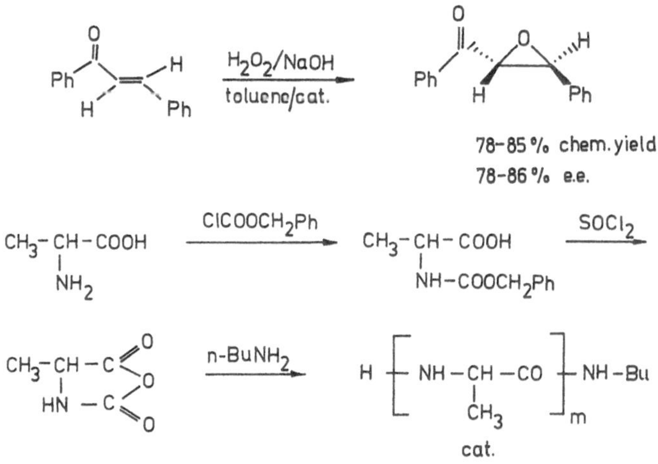
H2O2/NaOH
toluene/cat.
78-85 % chem. yield
78-86 % e.e.
CH3-CH-COOH
NH2
ClCOOCH2Ph
CH3-CH-COOH
NH-COOCH2Ph
SOCl2
n-BuNH2
H [NH-CH-CO] NH-Bu
CH3
m
cat.

Scheme 4

Recently Flisak et al. [7] achieved an extension of this reaction – using commercially available poly (L)-leucine they succeeded in epoxidizing substituted chalkones with high enantioselectivity according to this method. The polymer catalyst can be recovered without any loss of activity and can be used for further asymmetric epoxidations.

2.2 Use of Chiral Molybdenum(VI) (oxo-diperoxo) Complexes

From hydrogen peroxide to the molybdenum peroxo complexes, which can be used as epoxidizing agents, it is only a small step because the latter can be conveniently prepared from molybdenum oxide and hydrogen peroxide.

Cooperation of the Kagan, Mimoun, and Schurig groups led to an optically active molybdenum complex being obtained with (S)-N,N-dimethyllactamide (DMLA) as ligand. The structure of this complex (see Scheme 5) was confirmed by X-ray investigations. Simple prochiral unsubstituted olefins such as trans-2-butene were epoxidized in good chemical yield and with the given enantiomeric excess [8].

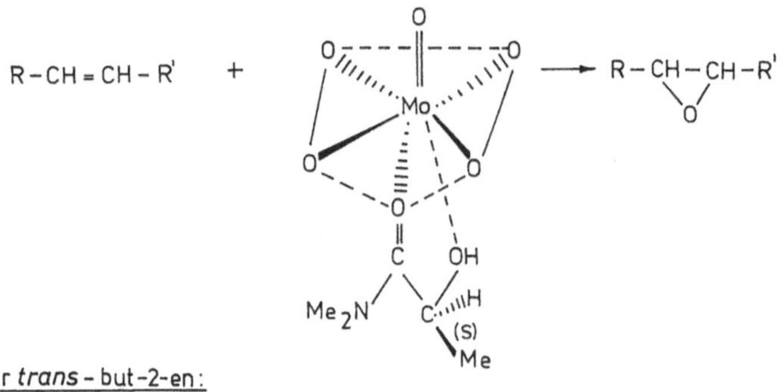

$$R-CH=CH-R' \quad + \qquad\qquad\qquad \longrightarrow \quad R-CH-CH-R'$$

for trans-but-2-en:

MoO(O₂)₂·DMLA	34.8 % e.e.
MoO(O₂)₂·PLA	55.0 % e.e.
MoO(O₂)₂·PLA − (2R,3R)-butandiol	89.8 % e.e.

Scheme 5

These investigations were continued, in particular by varying the radicals at the amidic nitrogen of the chiral ligand. With the Mo-complex of the (S)-piperidine lactamide (PLA) and with the addition of equimolar amounts of other additives the e.e. of the epoxide formed was increased to 90% [9].

2.3 Use of Chiral Dioxiranes

Recently, dioxiranes, which are available in solution by the reaction of ketones with monoperoxy sulfate, were demonstrated to be selective and efficient oxidizing agents [10]. Such diluted solutions, e.g. of dimethyldioxirane which can be stored at −20 °C, can be used for the epoxidation of various olefins, too. If chiral ketones are employed for the generation of dioxirane solutions, an optical induction should be expected after the epoxidation of the olefins.

Utilizing the optically active ketones which are shown in Scheme 6 as precursors, e.e.'s of up to 20% were obtained. The advantage of this method using in-situ generated dioxiranes, is that only catalytic amounts of optically active ketones are necessary [11].

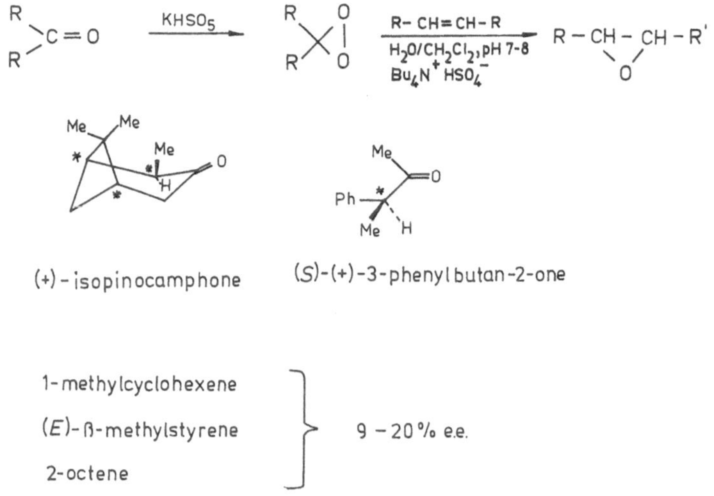
KHSO5
R–CH=CH–R
H2O/CH2Cl2, pH 7-8
Bu4N+ HSO4−
R – CH – CH – R'
(+)-isopinocamphone
(S)-(+)-3-phenylbutan-2-one
1-methylcyclohexene
(E)-β-methylstyrene
2-octene
9 – 20 % e.e.

Scheme 6

2.4 Use of *t*-Butyl Hydroperoxide in the Presence of Chiral Transition Metal Complexes

As is generally known, a very selective method for the formation of epoxides is the reaction of olefins with organic hydroperoxides in the presence of molybdenum, vanadium, or titanium compounds. This reaction is the basis of a new industrial route for the production of propylene oxide. At the end of the 1970s we investigated the epoxidation of long-chain olefins with *t*-butyl hydroperoxide (TBHP) in the presence of soluble and heterogenized Mo-catalysts. In connection with these investigations we epoxidized some olefins in the presence of V- and Mo-complexes, which contained (+)-3-trifluoroacetylcamphor as an optically active ligand [12].

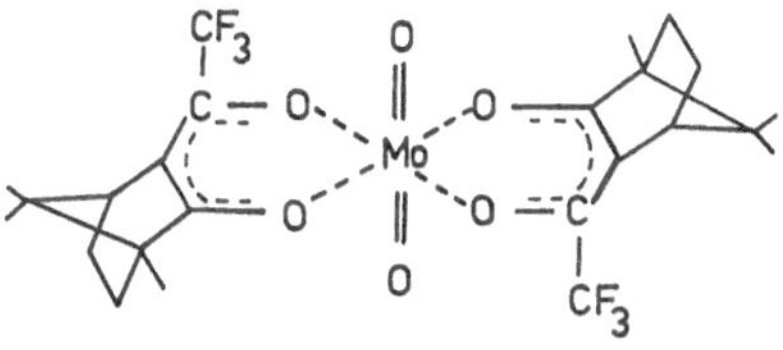

Olefine	Time (h)	HP-Conv. (%)	Selectivity (%)
1-octene	2	98	100
styrene	1	78	82
allyl chloride	2	98	82

Scheme 7

The chemical yield of epoxides was excellent, but we did not find any optical induction (Scheme 7). We discussed this result supposing that the ligand was oxidized and split off before the actual epoxidation step took place. It is now known that in the late 1970s similar investigations on Mo- and V-complexes with 1,3-diketones as optically active ligands were performed by various groups and, at the best, only low optical inductions were observed.

3 Enantioselective Epoxidation of Allylic Alcohols (Sharpless Reaction)

3.1 General Aspects

An actual breakthrough in the field of asymmetric epoxidation was the discovery of a new method for the preparation of homochiral epoxy alcohols by Katsuki and Sharpless in 1980 [13]. The main idea was to change the catalytic system and to use titanium tetraisopropoxide and L-(+)- or D-(−)-diethyl tartrate (DET). With water-free solutions of TBHP various allylic alcohols can be epoxidized with high enantioselectivity.

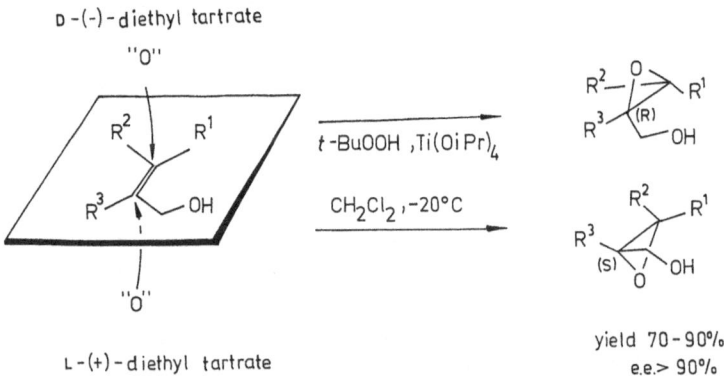

D-(-)-diethyl tartrate

"O"

t-BuOOH ,Ti(OiPr)$_4$

CH$_2$Cl$_2$,−20°C

L-(+)-diethyl tartrate

yield 70−90%
e.e.> 90%

Scheme 8

Some major advantages were clear from the very beginning:
— Upon using a given tartrate enantiomer the system seems forced to deliver the epoxide oxygen from the same enantioface of the olefin, regardless of the substitution pattern. When the olefinic unit is in the plane of the drawing as shown in Scheme 8, the use of (+)-DET leads to the addition of the epoxide oxygen from the bottom. Of course, when (−)-DET is employed, the epoxide oxygen is added from the top.
— The reaction gives evenly high asymmetric inductions throughout a large range of differently substituted allylic alcohols.
— All the starting compounds are commercially available and low-priced.

Some initial deficiencies of the method have been eliminated in the meantime by Sharpless and his coworkers. First, stoichiometric amounts of the catalytic system and TBHP were employed per mol of allylic alcohol, so it was not clear whether the reaction proceeded in a catalytic way. A few years ago it was found that, in the presence of molecular sieves, the amount of the catalytic system can be decreased to 5−6 mol-% of the allylic alcohol [14]. A comparison between the stoichiometric and catalytic variant was given for the epoxidation of oct-2-en-1-ol investigated by the Upjohn Company on a kilogram scale (Scheme 9) [15].

Comparison of the "Stoichiometric" and "Catalytic" Process.

	Stoichiometric	Catalytic
Allylic alcohol	8 Kg (62.5 mol)	18.8 Kg (147 mol)
Ti(Oi-Pr)$_4$	18.5 Kg (65.1 mol)	4.18 Kg (14.7 mol)
(+)-DET	14.3 Kg (69.4 mol)	3.34 Kg (16.16 mol)
TBHP (in toluene, 3.5 M)	35.5 L (124.2 mol)	63.0 L (220.5 mol)
4A° molecular sieves	—	3.76 Kg
CH$_2$Cl$_2$	500 L	200 L
Yield of 2**	7.2 Kg (80%)	17.1 Kg (81%)
Optical purity of 2**	> 98% ee	> 98% ee

* Dr. Joseph Timko, The Upjohn Company, unpublished results
** Following recrystallization

Scheme 9

The most significant advances of this modification are the use of smaller amounts of catalyst, a simplified workup procedure, and the possibility of working at much higher concentrations than in the original stoichiometric version. From Scheme 9, it follows that the stoichiometric run, although carried out on less than half the scale, required 2.5 times more solvent than the catalytic run.

A further deficiency of the method was the fact that only poor yields are obtained when the epoxy alcohol has a good water solubility. For example, after epoxidation of allyl and crotyl alcohol only 10–30% of the intact epoxy alcohol could be obtained by extraction. In the meantime, the workup has been improved and the water-soluble epoxy alcohols have been transformed in situ by nucleophilic reagents into the target products without isolation [14].

A new variant of the Sharpless epoxidation is a one-pot procedure. It is known that on photochemical oxidation of olefins in the presence of tetraphenylporphine the corresponding allyl hydroperoxides are formed. The latter were partly reduced into allylic alcohols, which were epoxidized by the remaining hydroperoxide in the presence of transition metal catalysts. By addition of L-(+)-DET to the photochemical oxidation of 2,3,3-trimethyl but-1-ene, Adam et al. [16] succeeded in preparing the (S)-epoxy alcohol in a good yield, while the epoxidation under standard conditions delivers only a smaller amount of product (Scheme 10).

Since the discovery of the asymmetric epoxidation reaction much information has been obtained about the mechanism of the process [17]. However this topic is still under discussion: In 1990, Corey [18] made a new proposal regarding the structure of a possible transition state and recently Sharpless et al. [19] began a new series of publications in this field. Therefore, the original literature should be referred to at this point.

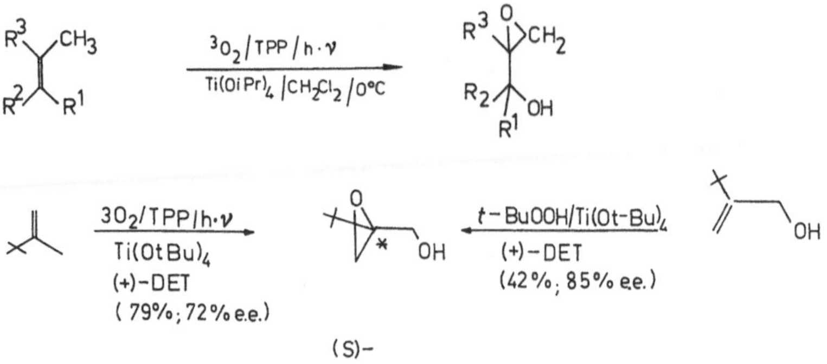

Scheme 10

Within the last ten years, a large number of papers have appeared in which the titanium/tartrate-catalyzed asymmetric epoxidation of allylic alcohols was used as a key step in the synthesis of carbohydrates, amino acids, pheromones, leucotrienes and other stereochemically homogeneous compounds. This development has been discussed in several reviews [20].

3.2 Epoxidation of C_5-Allylic Alcohols

Owing to our interest in chiral synthons as C_5-epoxy alcohols which are valuable building blocks for vitamins and other biologically active compounds, we investigated the enantioselective epoxidation of C_5-allylic alcohols and tried to transform the products obtained into other homochiral C_5-synthons. Until now, such oxygen-functionalized C_5-compounds have been prepared starting from natural products by chiral-pool syntheses.

We began our work with isoprene alcohol (see Scheme 11). While the Prilezhaev reaction with monoperoxyphthalic acid provided the appropriate dimethyl glycidol in good yield, we failed in our attempts to prepare the corresponding chiral epoxide by Sharpless epoxidation. Obviously, the reaction does not work in the case of this tertiary allylic alcohol. When the solution is allowed to warm up to room temperature a slow epoxidation takes place, however, the isolated dimethyl glycidol does not show any optical activity. Very recently the asymmetric epoxidation of a tertiary allylic alcohol in reasonable optical and chemical yield has been described [21].

Then we changed over to the isomer allylic alcohol, to 3-methyl-2-buten-1-ol (prenol). Being a primary alcohol, it was smoothly epoxidized under both stoichiometric and catalytic Sharpless conditions. While the stoichiometric method provides only moderate yields as the dimethyl glycidol is fairly watersoluble, the catalytic method affords the double yield. The e.e. amounts to 90% in both cases. Optical purity and e.e. of the 3,3-dimethyl glycidol were determined by polarimetry and ^{1}H-NMR in the presence of chiral europium shift-reagent [22].

On alkaline treatment of the 3,3-dimethyl glycidol, Payne rearrangement takes place. In alkaline aqueous solutions, 2,3-epoxy alcohols are in equilibrium with the corresponding 1,2-epoxy alcohols. Payne was the first to observe this

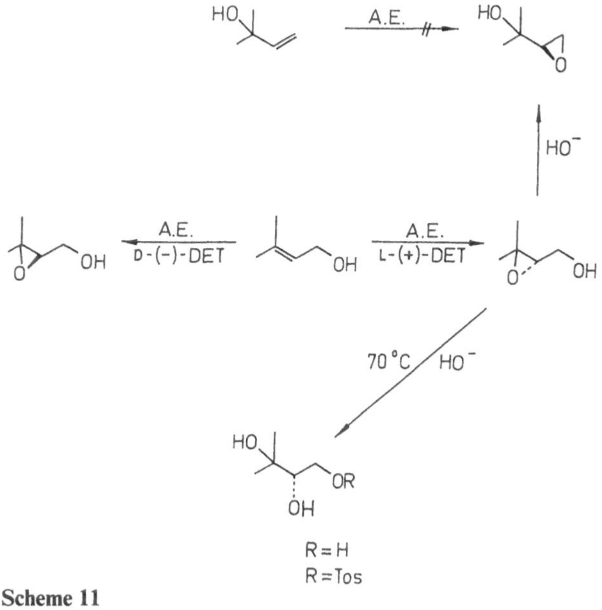

Scheme 11

transformation. The ratio of epoxy alcohols is dependent on the substrate, but in general the equilibrium is biased towards primary alcohols. That applies also to this case. Moreover, under particular conditions we succeeded in enriching the *t*-epoxy alcohol and isolating it by means of preparative gas chromatography. By optical rotation measurement and the e.e., which was determined by complexing gas chromatography [22], it was confirmed that the rearrangement proceeds stereospecifically with inversion at C-2. In this way we obtained the homochiral epoxide which we had failed to obtain by Sharpless epoxidation of isoprene alcohol.

After treatment of the primary epoxy alcohol with aqueous sodium hydroxide solution for several hours at 70 °C the (S)-triol was obtained unexpectedly. Obviously, the reaction takes place with retention and by ring-opening at C-3. Generally, the ring-opening reaction is expected on the sterically less hindered C-atom. From this result it appears that the base-catalyzed ring-opening of 2,3-epoxy alcohols is subjected to electronic factors in addition to steric influence. The (S)- triol can be transformed into the corresponding tosylate, which has a suitable leaving group for the coupling of this synthon to other molecules.

3.3 Use of Various Hydroperoxides in the Epoxidation of Prenol

Up to now in most cases, the commercially available TBHP or cumene hydroperoxide have been used in the asymmetric epoxidation reaction. Finn and Sharpless [23] reported that hydroperoxides other than TBHP can also be used and that secondary hydroperoxides, such as phenylethyl and cyclohexyl hydroperoxide, are

effective too. As far as we know, systematic investigations have not been reported on the influence of the structure of hydroperoxide on Sharpless epoxidation. Therefore we have studied the influence of the different hydroperoxides in the catalytic asymmetric epoxidation of prenol. The results obtained are given in Table 1.

Table 1. Catalytic asymmetric epoxidation of 3-methyl-2-buten-1-ol with different hydroperoxides[a]

hydroperoxide	yield (%)	e.e.[d] (%)
t-butyl	45–50[b]	90
cumyl	20[c]	72
1-methyl-1-phenylpropyl	19[c]	74
1-methyl-1-phenylbutyl		72
1-phenylethyl	14[c]	28
tetralyl-1-	35[b]	14

[a] All reactions were carried out at $-20\ ^\circ$C with L-(+)-DIPT/ Ti(OiPr)$_4$ as catalyst. 4A molecular sieves are used. The yields reported are isolated yields.
[b] Distilled epoxide.
[c] Purified by preparative gas chromatography.
[d] Enantioselectivity was determined by polarimetry and ^{1}H-NMR shift analysis.

It has been found that the most useful hydroperoxide in the epoxidation of prenol is TBHP. Thus we were able to isolate the homochiral 3,3-dimethyl glycidol with up to 50% yield after distillation. The use of cumene hydroperoxide, 1-methyl-1-phenylpropyl and 1-methyl-1-phenylbutyl hydroperoxide leads to some loss in enantioselectivity. The workup is more difficult than with TBHP. The chemical yield, although not optimized, is low and the reaction product was isolated only by preparative gas chromatography in these cases. Note that 1-phenylethyl hydroperoxide is not efficient for the asymmetric epoxidation of prenol. The e.e. obtained is significantly lower in comparison with all the tertiary hydroperoxides used. The application of tetralyl-1-hydroperoxide leads to sufficient yields but, on the other hand, to a very low optical induction. From the experimental data presented it becomes evident that it is very important to use bulky tertiary hydroperoxides in the asymmetric epoxidation to obtain a high enantiomeric excess. These results are in accordance with the mechanistic explanation of the Sharpless epoxidation which has been given recently by Corey [18]. The findings were only obtained in the case of one substrate and further investigations are necessary.

3.4 Kinetic Resolution of Allylic Alcohols and Chiral Hydroperoxides

In addition to the preparation of optically active epoxide intermediates, the titanium/tartrate-catalyzed asymmetric epoxidation can be used to resolve racemic

E. Höft

allylic alcohols. Upon adding half of the required amount of TBHP, one enantiomer is consumed due to the difference in the reaction rate, while the other remains unchanged. The enantiomer selected for epoxidation depends on whether a D-(−)- or L-(+)-tartrate is used. This method of kinetic resolution has been proved on a multitude of secondary allylic alcohols. One example of our investigation may elucidate this method (Scheme 12). Oct-1-en-3-ol was epoxidized with TBHP in the molar ratio 2:1 in the presence of titanium isopropoxide and one of the diisopropyl tartrate (DIPT) enantiomers.

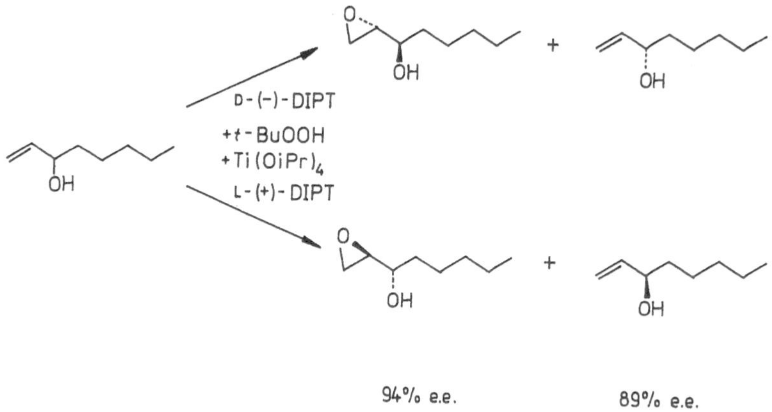

D-(−)-DIPT
+t-BuOOH
+Ti(OiPr)₄
L-(+)-DIPT

94% e.e.   89% e.e.

Scheme 12

Besides the corresponding epoxy alcohol the enantiomer of the allylic alcohol reacting more slowly was recovered with high e.e. In this case, the ratio of enantiomers was determined similarly to König method [24] by gas chromatography on a column with perpentylated α-cyclodextrine.

Analogously, we have tried to use the titanium/tartrate-catalyzed asymmetric epoxidation for the kinetic resolution of the other component of the reaction, namely the hydroperoxide when it is chiral. If the molar ratio of allylic alcohol and hydroperoxide is 1:2 and the reaction is quenched after the consumption at half the amount of hydroperoxide, in the ideal case the homochiral epoxy alcohol, the homochiral hydroperoxide and the homochiral carbinol should be expected.

The first preliminary results with the chiral 1-methyl-1-phenylpropyl hydroperoxide show (see Table 2) that, in principle, the expected stereo-differentiation occurs.

The hydroperoxide isolated after epoxidation shows optical activity. However, the polarimetric data measured were difficult to reproduce and, moreover, the rotation data for the pure hydroperoxide enantiomer are not known because, as yet, it has not been isolated. Therefore, we determined the ratio of hydroperoxide enantiomers according to a method we have recently elaborated. In the case of prenol the e.e. is only indicated, on geraniol it is a little bit higher and using cinnamyl alcohol it reaches about 12% both in the stoichiometric and in the catalytic modification of the reaction. Of course, these results are not applicable for preparative purposes and, therefore, we intend to continue these investigations

74

Table 2. Kinetic resolution of 1-methyl-1-phenylpropyl hydroperoxide

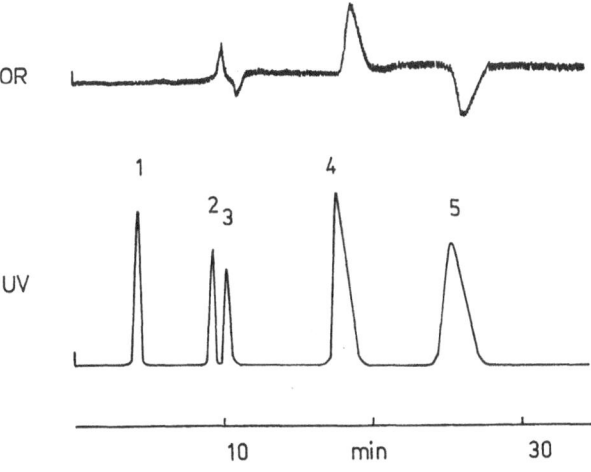

allylic alcohol	R^1	R^2	DIPT	ratio of enantiomers (%)		e.e. (%)
				(+)	(−)	
prenol	CH_3	CH_3	L−(+)	52	48	4
			D−(−)	48	52	4
geraniol	CH_3	$(CH_3)_2C=CH$ $(CH_2)_2$	L−(+)	53	47	6
			D−(−)	47	53	6
cinnamyl alcohol	H	C_6H_5	L−(+)	45	55	10
			D−(−)	56	44	12

by varying the allylic alcohol and the chiral hydroperoxide as the hitherto known methods for the preparation of homochiral hydroperoxides are particularly complicated and uncertain concerning the stereoselectivity.

In the following, let me return to the determination of the ratio of enantiomers in chiral hydroperoxides. As the gas-chromatographic methods are not very suitable due to the thermal sensitivity of hydroperoxides we tried to use ^{1}H-NMR and HPLC methods without derivatization of the hydroperoxide. While satisfacto-

Fig. 1. Optical resolution of partly decomposed 1-methyl-1-phenylpropyl hydroperoxide. Eluent: water-saturated hexane/2-propanol, 98/2, v/v; upper: optical rotation; lower: UV; 1 unknown, 2 (+)-enantiomer and 3 (−)-enantiomer of the related alcohol, 4 (+)-enantiomer and 5 (−)-enantiomer of the hydroperoxide

ry conditions were not found with the help of ^{1}H-NMR and chiral shift reagents we succeeded in separating the enantiomers by HPLC on a chiral stationary phase consisting of tris(3,5-dimethylphenyl carbamate) coated on macroporous silica (CHIRALCEL OD).

The chromatograms of the racemic hydroperoxide which was partly decomposed into the corresponding carbinol were recorded by UV and a chiral detector (Fig. 1). First the carbinol was separated and then a better optical resolution and longer retenton time was observed for the hydroperoxide. A comparison of both chromatograms indicates that the optical rotation of the enantiomer of the carbinol, which was eluted first, was plus. This method was successfully used for the enantiomeric separation of other chiral hydroperoxides on an analytical scale [25]. Thus for the first time a practical method is available for the determination of enantiomers of hydroperoxides and related alcohols in the same solution.

Summing up, a short review about the methods of enantioselective epoxidation with peroxidic reagents has been given. In the forefront of the discussion there was the titanium/tartrate-catalyzed asymmetric epoxidation discovered by Katsuki and Sharpless, which is characterized by simplicity, reliability, and high enantio-selectivity, but also by well-defined limits. Moreover, I have tried to include some results obtained in this field by our group adding them to the knowledge of asymmetric epoxidation.

I would like to thank my coworkers Dr. H. J. Hamann and Dr. L. Rüffer for fruitful cooperation and our colleagues from the analytical department of our institute Dr. B. Costisella, Dr. W. Hiller, Dr. A. Kunath, and Dr. T. Reiher for their help.

5 References

1. Ewins RC, Henbest HB, McKervey MA (1967) JCS Chem Commun 1085;
   Montanari F, Moretti I, Torre G (1968) Boll Sci Fak Chim Ind Bologna 113;
   Pirkle WH, Rinaldi PL (1977) J Org Chem 42: 2080
2. Payne GB, Deming PH, Williams PH (1961) J Org Chem 26: 659
3. Ben Hassine B, Gorsane M, Geerts-Evrard F, Pecher J, Martin RH, Castelet D (1986) Bull Soc Chim Belg 95: 547
4. Helder R, Hummelen JC, Laane RWPM, Wiering JS, Wynberg H (1976) Tetrahedron Lett 1831
5. Pluim H, Wynberg H (1980) J Org Chem 45: 2498
6. Juliá S, Masana J, Vega JC (1980) Angew Chem 92: 968;
   Juliá S, Guixer JG, Masana J, Rocas J, Colonna S, Annuziata R, Molinari H (1982) JCS Perkin Trans I: 1317
7. Baures PW, Eggleston DS, Flisak JR, Gombatz K, Lantos I, Mendelson W, Remich JJ (1990) Tetrahedron Lett 31: 6501
8. Kagan HB, Mimoun H, Mark C, Schurig V (1979) Angew Chem 91: 511
9. Schurig V., Hintzer K, Leyrer U, Mark C, Pitchen P, Kagan HB (1989) J Organomet Chem 370: 81
10. Adam W, Curci R, Edwards JO (1989) Acc Chem Res 22: 205
11. Curci R, Fiorentino M, Serio MR (1984) JCS Chem Commun 155
12. Döbler Ch, Höft E (1978) Z Chem 18: 218
13. Katsuki T, Sharpless KB (1980) J Am Chem Soc 102: 5974

14. Gao Y, Hanson RM, Klunder JM, Ko SY, Masamune H, Sharpless KB (1987) J Am Chem Soc 109: 5765
15. Sharpless KB (1988) Janssen Chim Acta 6: 3
16. Adam W, Griesbeck A, Staab E (1986) Tetrahedron Lett 27: 2839
17. Finn MG, Sharpless KB (1985) In: Morrison JD (ed) Asymmetric synthesis, vol 5, Academic, New York, p 247
18. Corey EJ (1990) J Org Chem 55: 1693
19. Woodard SS, Finn MG, Sharpless KB (1991) J Am Chem Soc 113: 106
    Finn MG, Sharpless KB (1991) J Am Chem Soc 113: 113
20. Rossiter BE (1985) In: Morrison JD (ed) Asymmetric synthesis, vol 5, Academic, New York, p 193
    Pfenninger A (1986) Synthesis 89
    Höft E, Hamann HJ (1987) Mitteilungsbl Chem Ges DDR 34: 10
21. Takano S, Iwabuchi Y, Ogasawara K (1991) Tetrahedron Lett 32: 3527
22. Höft E, Hamann HJ, Rüffer L, Jáky M (1991) In: Simándi LI (ed) Dioxygen activation and homogeneous catalytic oxidation, Elsevier, Amsterdam, p 537
23. Finn MG, Sharpless KB (1985) In: Morrison JD (ed) Asymmetric synthesis, vol 5, Academic, New York, p 267
24. König, WA (1989) Nachr Chem Tech Lab 37: 471
25. Kunath A, Höft E, Hamann HJ, Wagner J (1991) J Chromatogr 588: 352

Synthesis of Dicarboxylic Acids by Transition-Metal Catalyzed Oxidative Cleavage of Terminal-Unsaturated Fatty Acids

Siegfried Warwel, Michael Sojka and Mark Rüsch gen. Klaas

Institute of Technical Chemistry and Petrochemistry, Aachen University of Technology, D-5100 Aachen, FRG

Table of Contents

Topics in Current Chemistry, Vol. 164
© Springer-Verlag Berlin Heidelberg 1993

Siegfried Warwel, Michael Sojka and Mark Rüsch gen. Klaas

9-Decenoic- and 13-tetradecenoic methylesters, obtained by Re-catalysed metathesis of natural C_{18}- and C_{22}-fatty esters with ethylene as well as the industrially produced 10-undecenoic methylester were used as starting materials for the preparation of dicarboxylic esters. Two different reaction routes were applied. Ketonisation of the terminal unsaturated fatty acid esters by Wacker-oxidation using $PdCl_2/CuCl$ or $RhCl_3/FeCl_3$ as catalysts led to methyl keto-fatty acid esters, which were oxidatively cleaved by Mn-catalysed oxidation with air at 115 °C to mixtures of C_8-/C_9-, C_9-/C_{10}- and C_{12}-/C_{13}-dicarboxylic monomethylesters with conversion rates and selectivities of 90%.

Pure C_9-, C_{10}- and C_{13}-dicarboxylic acids were achieved in isolated yields of 80–90% by an one step oxidative cleavage of the terminal unsaturated fatty acids with peracetic acid as oxidant and different ruthenium compounds e.g. $Ru(acac)_3$ as catalysts. Starting with 10-undecenoic acid, a simple catalyst recycling could be established.

Replacing peracetic acid with H_2O_2 or acetic acid/H_2O_2 the ruthenium-catalyzed oxidative cleavage of $C=C$-bonds did not proceed because of an unproductive decomposition of H_2O_2. With Re_2O_7 as catalyst hardly any decomposition of H_2O_2 took place and using 1,4-dioxane as solvent, olefins were converted to vicinal diols in fair yields.

1 Introduction

In the industrial oleochemistry, oxidations of unsaturated fatty substances are limited to epoxidation and − to a much lower degree − to ozonolysis yielding mono- and dicarboxylic acids [1]. Up till now, metal-complex catalysed oxidations have not been applied.

Ozonolysis is an elegant method for the oxidative cleavage of $C=C$-bonds [2, 3]. It is applied in fat-chemistry to produce azelaic (nonanedioic) acid and pelargonic (nonanoic) acid [1, 4]:

$$CH_3-(CH_2)_7-CH=CH-(CH_2)_7-COOH$$

$$\downarrow +O_3$$

$$CH_3-(CH_2)_7-CH\begin{array}{c} O \\ \diagdown \diagup \\ O-O \end{array}CH-(CH_2)_7-COOH$$

$$\downarrow +O_2$$

$$CH_3-(CH_2)_7-COOH + HOOC-(CH_2)_7-COOH$$

Analogously erucic (13-docosenoic) acid may be converted to brassylic (tridecane-dioic) acid and pelargonic acid:

$$C_8H_{17}-CH=CH-(CH_2)_{11}-COOH \xrightarrow{+O_3/O_2} \begin{array}{c} HOOC-(CH_2)_{11}-COOH \\ + \\ C_8H_{17}-COOH \end{array}$$

Aldehydes and their derivates are produced by reductive working-up of the primary formed ozonides.

Though ozonolysis is very selective and high yielding, the technical use is, according to industrial complaints [5], difficult for reasons of economy and safety. The development of ozone-free processes for oxidative $C=C$-cleavage is therefore a serious task in fat-chemistry research.

At this point we started our own reseach studying selective oxidations of $C=C$-bonds in unsaturated fatty matters catalysed by particular transition-metal compounds.

Regarding the ozone-free oxidative $C=C$-cleavage the following concepts were taken into account:
a) an indirect, two-step $C=C$-cleavage, characterized by the reaction-sequence
   unsaturated fatty acid → keto acid → carboxylic/dicarboxylic acid
b) the direct ruthenium-catalysed $C=C$-cleavage by percarboxylic acids

2 Oleochemicals as Starting Materials for the Synthesis of Dicarboxylic Acids

As oleochemicals for the preparation of dicarboxylic acids we did not choose natural C_{18}- or C_{22}-fatty acids with an internal double bond (oleic-, erucic acid),

as they were fed in ozonolysis, but terminal unsaturated $C_{10}-C_{14}$-fatty acids or their methylesters, especially

> 9-decenoic acid methylester
> 10-undecenoic acid methylester
> 13-tetradecenoic acid methylester.

We preferred these educts for the following reasons:

First we wanted to avoid the formation of a middle chain monocarboxylic acid (pelargonic acid) as a by-product. Second oleic acid or its methylester are, not industrially available in pure form. Produced by hydrolysis or transesterification of tallow or vegetable oil, they are always contaminated by other C_{18}-fatty acids/methylesters, as for example linoleic acid/methylester, which cannot be separated in an economic manner.

In contrast, the $C_{10}-C_{14}$-fatty acids or their methylesters, which we used, are available in absolutely pure qualities.

10-Undecenoic acid is an industrial product, which is produced by the pyrolysis of ricinolic acid methylester, based on castor oil [1].

To prepare 9-decenoic and 13-tetradecenoic acid methylester, we used the transition-metal catalysed olefin metathesis [6, 7], actually applied in petro- and polymerchemistry in seven industrial processes [8]. As stated first by C. Boelhouwer et al. [9, 10], this reaction is valid not only for olefins, but for unsaturated fatty acid esters, too.

Admittedly — in contrast to the metathesis of unfunctionalized olefins — much higher concentrations of the catalysts $WCl_6 + SnR_4$ [9] and $Re_2O_7/Al_2O_3 + SnR_4$ [10], which are suitable for ester metathesis, were necessary. We succeeded in developing new catalysts — $Re_2O_7 - B_2O_3/Al_2O_3 - SiO_2 + Sn(n-C_4H_9)_4$ —, whose activity dramatically outperform the common systems [12–14]. The choice of the correct supporting material   amorphous alumosilicates, manufacturated by Condea Chemie GmbH, Brunsbüttel, Germany, with a SiO_2-percentage of 40 weight-% proved to be the best — and modification with B_2O_3 was also essential. In addition, the very important newly developed tin-alkyl-free catalysts $CH_3ReO_3 + Al_2O_3$ and $CH_3ReO_3 + Al_2O_3 - SiO_2$ with 87 weight-% SiO_2 ("Herrmann's catalyst") [15–18] gain a new dimension of activity, if $B_2O_3 - Al_2O_3 - SiO_2$ is used as support/co-catalyst [19].

By using our new Re_2O_7-catalysts we were able to prepare 9-decenoic acid methylester not only by metathetic ethenolysis of pure oleic acid methylester according to

$$C_8H_{17}-CH=CH-(CH_2)_7-COOCH_3 \; \underset{\longleftarrow}{\overset{[cat]}{\longrightarrow}}$$
$$+$$
$$CH_2=CH_2$$

$$C_8H_{17}-\underset{\underset{CH_2}{\|}}{CH} \quad \underset{\underset{CH_2}{\|}}{CH}-(CH_2)_7-COOCH_3$$
$$+$$

but even by using fatty acid methylester mixtures, which are available in industrial scale by transesterification of tallow or vegetable oils (rapeseed-, soybean-sunflower oil). Fatty acid methylesters from "high oleic" sunflower oil proved to be especially suitable for producing 9-decenoic acid [13] (Table 1).

Table 1. 9-Decenoic acid methylester by metathesis of fatty acid methylesters with ethylene

Methylester from	conversion (%)	Methyl 9-decenoate	
		yield (% of the theoretical; isolated)	purity (%, GC)
oleic acid	94	83	98
sunflower oil ($C_{18:1}$)	83	75	97
rapeseed oil	80	61	97
tallow	83	59	97
sunflower oil ($C_{18:2}$)	83	64	97
soybean oil	83	60	97

Cat. $B_2O_3 - Re_2O_7/Al_2O_3 - SiO_2 + Sn(n-C_4H_9)_4$
$Re_2O_7/SnR_4/ester = 1/1.5/400$ (molar)
400 ml autoclave; t = 25 °C, 20 h, 50 bar ethylene

Under identical conditions erucic acid methylester (from erucic acid rich rapeseed oil) was converted to 13-tetradecenoic methylester by metathetic ethenolysis.

$$C_8H_{17}-CH=CH-(CH_2)_{11}-COOCH_3 \underset{\longleftarrow}{\overset{[cat.]}{\longrightarrow}}$$
$$+$$
$$CH_2=CH_2$$

$$C_8H_{17}-\underset{\parallel}{CH} \quad \underset{\parallel}{CH}-(CH_2)_{11}-COOCH_3$$
$$\underset{CH_2}{\quad} + \underset{CH_2}{\quad}$$

With a conversion of 91% the C_{14}-ester was isolated in a yield of 77% and a purity of 97% (according to GC).

3 Synthesis of Dicarboxylic Acids from Unsaturated Fatty Acids via Palladium-Catalysed Ketonisation [20]

3.1 Synthesis of Keto-Fatty Acids

Keto-fatty acids may be prepared conventionally by epoxidation of unsaturated fatty acids and isomerisation of the prepared epoxy fatty acids [21]. In contrast, we directly oxidized the unsaturated fatty acids to methylketo-fatty acids.

We first chose the well-known system $PdCl_2/CuCl_2$ [22, 23] as catalyst, which is used industrially for the oxidation of ethylene to acetaldehyde (Wacker-Hoechst-Process). The reaction is based on the observation, that ethylene is oxidized to acetaldehyde by an aqueous solution of $PdCl_2$ (F. C. Philips, 1884); simultaneously a precipitation of metallic palladium occurs. The precipitated palladium can be reoxidized by addition of Cu^{2+}-salts, and by passing air through the solution to convert Cu^+ to Cu^{2+} the reaction becomes catalytic. The oxygen, inserted in the organic molecule does not originate from the oxidant but from the water.

$$CH_2=CH_2 + PdCl_2 + H_2O \longrightarrow CH_3CHO + Pd + 2\,HCl$$
$$Pd + 2\,CuCl_2 \longrightarrow PdCl_2 + 2\,CuCl$$
$$2\,CuCl + 1/2\,O_2 + 2\,HCl \longrightarrow 2\,CuCl_2 + H_2O$$
$$\overline{CH_2=CH_2 + 1/2\,O_2 \longrightarrow CH_3CHO}$$

The ketonisation of higher olefins is possible without any problems, if the $C=C$-bond is a terminal one; so 1-octene can be converted to 2-octanon and 10-undecenoic acid to 10-oxo-undecanoic acid [24, 25]. The direct ketonisation of 9-decenoic acid, however, not prepared by olefin metathesis but by Barbier-Wieland-degradation of 10-undecenoic acid, has been reported, too [26].

Our own experiments proved the modified catalyst system $PdCl_2/CuCl$ to be advantageous. That way, the Cu^{2+} necessary to reoxidize the metallic palladium, was not added in the form of $CuCl_2$, but was formed during the reaction.

The reaction is highly influenced by the solvent; therefore among others NMP, DMF, dioxane, methyl formamide, dimethyl acetamide and tetramethyl urea were tested. The best results were achieved in DMF and tetramethyl urea. Using these solvents and $PdCl_2/CuCl$ or $RhCl_3/FeCl_3$ [27] as catalysts, oxidations of 10-undecenoic acid methylester, 9-decenoic acid methylester and 13-tetradecenoic acid methylester were carried out on a preparative scale (Table 2).

Under ideal conditions the $C_{10} - C_{14}$ fatty acids were converted, palladium-catalysed, to 100%; the isolated yields of 9-oxo-decanoic, 10-oxo-undecanoic and 13-oxo-tetradecanoic acid methylester were about 70%. $RhCl_3/FeCl_3$ as catalyst was more selective but the yields were lower.

Table 2. Ketonisation of ω-unsaturated fatty acid esters with $PdCl_2/CuCl$ or $RhCl_3/FeCl_3$

n	educt	catalyst	solvent	C*	Y**	S***
7	9-Decenoic acid ME	$PdCl_2/CuCl$	DMF/H_2O	85	65	92.4
8	10-Undecenoic Acid ME	$PdCl_2/CuCl$	DMF/H_2O	85	66	96.3
11	13-Tetradecenoic Acid ME	$PdCl_2/CuCl$	DMF/H_2O	80	61	93.1
7	9-Decenoic acid ME	$PdCl_2/CuCl$	TMH/H_2O	100	67	97.7
8	10-Undecenoic Acid ME	$PdCl_2/CuCl$	TMH/H_2O	99	73	96.9
11	13-Tetradecenoic Acid ME	$PdCl_2/CuCl$	TMH/H_2O	99	71	94.5
7	9-Decenoic acid ME	$RhCl_3/FeCl_3$	DMF/H_2O	69	41	100
8	10-Undecenoic Acid ME	$RhCl_3/FeCl_3$	DMF/H_2O	75	57	100

  * conversion according to GC-analysis (%)
 ** isolated yield (%)
*** selectivity
$PdCl_2/CuCl/ester = 1/10/10$ molar; 20 °C, 17 h, 1 bar O_2
$RhCl_3/FeCl_3/ester = 1/2/12$ molar; 80 °C, 24 h, 1 bar O_2

The selectivity of the palladium catalysed ketonisation to the desired methylketo-carboxylic acids is strongly dependent on the solvent and the reaction temperature. While at room temperature selectivities were 90% or above, at 60 °C in DMF/H_2O it was about 80%. The influence of the solvent is shown most clearly by replacing DMF with dioxane. In dioxane the oxidation of 10-undecenoic acid methylester, performed with a selectivity of 55%, at 60 °C, was a mere 45% (cf. Table 3). Isomeric oxo-undecanoic acid methylesters (9-, 8-, 7-, 6-, 5-, 4-) were detected by GC/MS.

Table 3. Ketonisation of 10-undecenoic acid methylester with $PdCl_2/CuCl$ — influence of solvent and temperature —

temp. °C	solvent	conversion (% acc. to GC)	selectivity (%) 10-oxo-undecanoic acid ME
rt	DMF/H_2O	85	96.3
rt	$dioxane/H_2O$	96	55.3
60	DMF/H_2O	96	82.2
60	$dioxane/H_2O$	100	45.3

molar ratio $PdCl_2/CuCl/ester = 1/10/10$; 17 h; 1 bar O_2
ME = methylester

3.2 Dicarboxylic Acid by Oxidative Cleavage of Keto-Fatty Acids

Ketones are cleaved by oxidation to carboxylic acids, which process is industrially applied to convert cyclohexanone to adipic (hexanedioic) acid and cyclododeca-none to dodecanedioic acid. These oxidations are carried out preferably vanadium-

Siegfried Warwel, Michael Sojka and Mark Rüsch gen. Klaas

catalysed by nitric acid, but may also be carried out in the presence of Cu- and Mn-acetates with air as the oxidant [28]. In fat chemistry 12-oxo-stearic acid methylester, which is accessible by rearrangement of ricinoleic acid methylester [29], was cleaved oxidatively [30].

First we examined the oxidative cleavage of open-chain aliphatic ketones (2-nonanone and 8-pentadecanone) by transition-metal compounds and air. At 100–120 °C the oxidative cleavage yielding carboxylic acids performed quantitatively. 2-Nonanone and 8-pentadecanone were converted to a mixture of caprylic (octanoic) and oenanthic (heptanoic) acid; i.e. the oxidative cleavage attacks — according to expectation — most preferentially the keto-group. A further degradation of the carbon-chain, which is typical for the common paraffin-oxidation, occurs only to a small degree — however, it could not be surpressed totally. Only manganese salts (chloride, acetate, acetylacetonate, stearate), used in concentrations of $1-2$ mol-% relating to the ketone, were found to be catalytically active. The anion was of no influence on the oxidative cleavage. Compounds of other metals (Ti, V, Cr, Fe, Co, Ni, Cu, Mo, Ru, Rh, Pd, Pt, W, Re, Ce, Yb) were not active.

After these experiments, concerning the oxidative cleavage of aliphatic ketone, the methyl-keto fatty acids, prepared by direct oxidation of ω-unsaturated fatty acids, were put to use. The oxidations were carried out in a bubble column reactor at 115 °C over a retention time of 60 min with 2 mol-% Mn(stearate)$_2$ as catalyst. Afterwards the reaction mixture was cooled to room temperature and the solvent acetic acid was removed at 30 °C in vacuum.

To analyse the reaction products with GC, quantitative esterification of the carboxyl-group to the corresponding methylesters by etheral diazomethane was required.

The results indicated, that the methylketo fatty acids had been cleaved mainly at the carbonyl-group to produce dicarboxylic acids, being one or two C-atoms shorter than the original acid:

$$CH_2=CH-(CH_2)_n-COOCH_3 + H_2O/O_2 \xrightarrow{[cat.]} CH_3-\overset{O}{\overset{\|}{C}}-(CH)_n-COOCH_3$$

$$CH_3-\overset{O}{\overset{\|}{C}}-(CH_2)_n-COOCH_3$$

$$+O_2 \Big|\ \substack{[Mn(stearat)_2] \\ 115\,°C/1\,h}$$

$$CH_3-COOH \qquad\qquad HCOOH$$
$$+ \qquad\qquad\qquad\qquad +$$
$$HOOC-(CH_2)_{n-1}-COOCH_3 \qquad HOOC-(CH_2)_n-COOCH_3$$

$n = $ 7: 9-oxo-decanoic acid $\longrightarrow$ cork-/acelaic acid (C_8/C_9)
$n = $ 8: 10-oxo-undecanoic acid $\longrightarrow$ acelaic-/sebacic acid (C_9/C_{10})
$n = $ 11: 13-oxo-tetradecanoic acid $\longrightarrow$ dodecanedioic-/brassylic acid (C_{12}/C_{13})

The molar ratio of longer to shorter dicarboxylic acid was always 2 to 1. This makes it clear, that the cleavage next to the carbonyl-group did not happen statistically but preferentially between the methyl- and the carbonyl-group. At conversions as high as 90%, these pairs of dicarboxylic acids amount to 90 mol-% of all dicarboxylic acids formed. The other 10 mol-% of the products are mainly dicarboxylic acids, being 3 or more C-atoms shorter than the educt.

Table 4. Oxidative cleavage of methyl-keto fatty acid esters

	9-decenoic acid methylester	10-undecenoic acid methylester	13-tetradecenoic acid methylester
Conversion[a]	90.8	86.3	84.2
Yield[b]	72.0	72.6	71.3
Selectivity[c]	91.2	88.9	89.0
composition of the resulting dicarboxylic acid monomethylesters (mol-%)			
C_{13}			60.2
C_{12}			28.8
C_{11}			5.2
C_{10}		58.3	1.9
C_9	58.2	30.8	1.1
C_8	33.0	6.7	1.0
C_7	5.2	2.5	traces
C_6	1.9	1.5	traces
C_5	1.0	1.0	traces
C_4	0.7	0.5	traces

[a] conversion acc. to GC-anaylsis (%)
[b] isolated yield (%) of all dicarboxylic acid monomethylesters
[c] selectivity (%, related to the two main products)
Charge: 0.016 mol oxo-fatty acid methylester; 0.32 mmol (0.2 g) Mn(stearate)$_2$; 115 °C; 1 h; $p_{O_2} = 1$ bar; 15 l/h O_2; 0.36 mol (21.6 g) acetic acid as solvent

These C_8-/C_9-, C_9-/C_{10}- and C_{12}-/C_{13}-dicarboxylic acids, pepared by oxidative cleavage of methyl-keto fatty acids are suitable for the production of plasticizers and ester based lubricants.

4 Synthesis of Dicarboxylic Acid from Unsaturated Fatty Acids by Ruthenium-Catalysed Oxidative Cleavage of C=C-Double Bonds

4.1 Introduction

Beside the two-step oxidative cleavage: unsaturated fatty acids → ketofatty acids → dicarboxylic acids, we also studied direct oxidative cleavage of C=C-bonds without ozone. Such an alternative to ozonolysis is given by the ruthenium catalysed oxidative C=C-cleavage. This synthesis is based on work of C. Djerassi

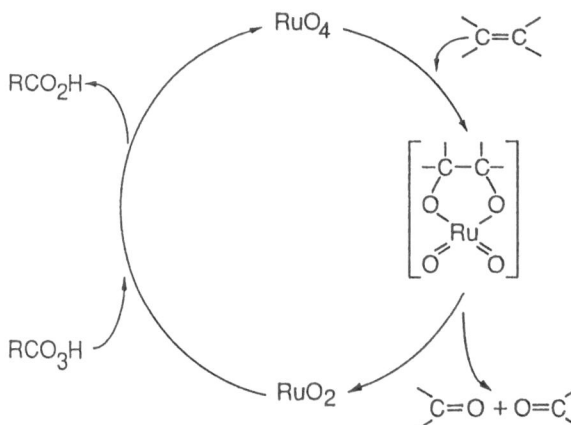

Fig. 1. Catalytic cycle of the ruthenium catalysed oxidative cleavage of C=C-bonds with percarboxylic acid

and R. Engle [31] concerning the oxidation potential of RuO_4 and of L. R. Berkovitz and P. N. Rylander [32], who cleaved oxidatively C=C-bonds by stoichiometric amounts of RuO_4. However this procedure is problematical regarding economy and toxicity. Therefore catalytic processes are much more important [reviews: 33–39]. Here ruthenium compounds such as $RuCl_3$, RuO_2 etc. are converted to ruthenium species in a high oxidation state (RuO_4), which afterwards attack C=C-bonds electrophilicly to form not isolable cyclic esters. Via ring-opening, carbonyl compounds and ruthenium in lower oxidation states are generated. Then the latter is reoxidized, so that the catalytic cycle is closed (Fig. 1).

As oxidants $NaIO_4$ [40, 41], NaOCl [5, 42–46], Cer^{4+}-compounds [47, 48] and peracetic acid [49–51] have been used.

4.2 Ruthenium-Catalysed Ocidative Cleavage of 1-Octene by Peracetic Acid-Influence of Various Parameters

Regarding the direct oxidative cleavage of terminal unsaturated fatty acids, the reaction was first studied systematically with 1-octene. The ruthenium-catalysed oxidative cleavage with peracetic acid produces oenanthic (heptanoic) acid and formic acid.

$$C_6H_{13}-CH=CH_2 \xrightarrow[\substack{+CH_3CO_3H \\ -CH_3CO_2H}]{[cat]} C_6H_{13}-COOH + HCOOH$$

Byproducts are heptanal, 1,2-octenoxid, 1,2-octanediol, mono- and diacetates of 1,2-octanediol and caproic (hexanoic) acid, due to oxidative degradation. Figure 2 clarifies the main parts of the complex reaction [cf. 52].

The oxidative cleavage of the olefin and especially the following oxidation of the aldehydes to carboxylic acids by peracetic acid are fast reactions. The side reaction (Prileshajev-reaction) to the epoxide proceeds much slower under the

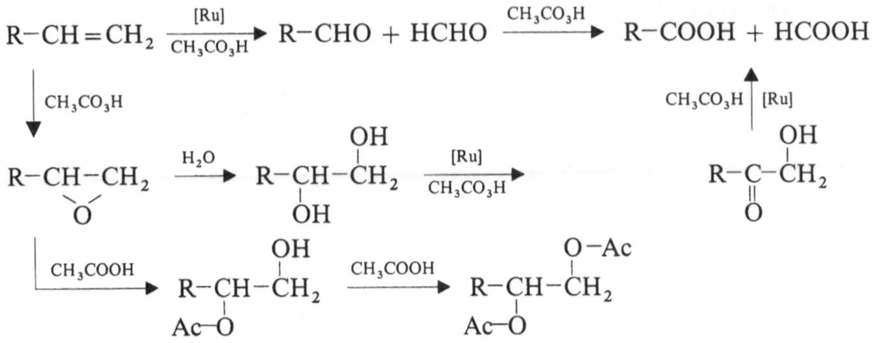

$$R-CH=CH_2 \xrightarrow[CH_3CO_3H]{[Ru]} R-CHO + HCHO \xrightarrow{CH_3CO_3H} R-COOH + HCOOH$$

Fig. 2. Ruthenium-catalysed oxidative cleavage of α-olefines — main and side reactions

applied conditions. If vicinal diols are formed, they will be cleaved like the olefins. In contrast to vicinal diols, their mono- and diacetates were not cleaved; we were able to establish this fact by experiment. If the best conditions for the reaction are chosen, all by-products together will amount to less than 10% (according to GC) of the reaction product.

Experiments to determine the best reaction conditions led to the following results:

Peracetic acid is the oxidant of choice; other percarboxylic acids as performic acid or m-CPBA proved to be less suitable, because they cause mainly epoxidation and consecutive reactions. Peracetic acid was used in the form of the so called "peracetic acid in equilibrium" or in the form of the distilled aqueous peracetic acid. "Peracetic acid in equilibrium" is made from acetic acid and H_2O_2 in the presence of H_2SO_4; distilled aqueous peracetic acid is produced from it by distillation as an azeotropic mixture of peracetic acid/water.

Table 5 shows the composition of both peracetic acid types. In contrast to "peracetic acid in equilibrium", which is indefinitely stable at 5 °C and may therefore be stored, distilled peracetic acid decomposes. For industrial purposes it is produced on site immediately before use. However, for reasons of safety, peracetic acid should not be distilled in the laboratory!

Table 5. Composition of various peracetic acids [53]

$$CH_3COOH + H_2O_2 \xrightleftharpoons{[H^+]} CH_3COOOH + H_2O$$

weight-%	CH₃COOOH	CH₃COOH	H₂O₂	H₂O	H₂SO₄
A	38	46	5	10	1
B	33	4	1	42	—

A: Peracetic acid in equilibrium. B: Distilled aqueous peracetic acid

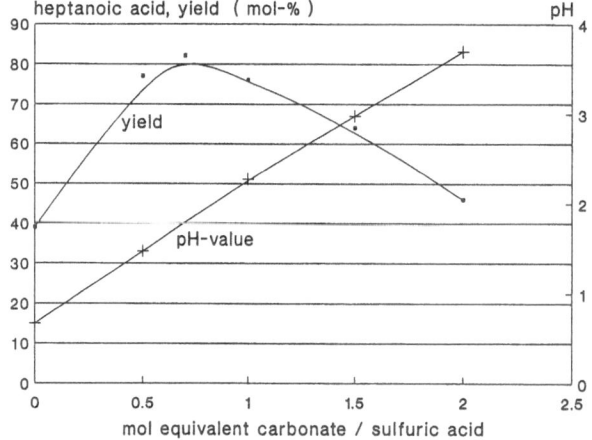

Fig. 3. Ruthenium-catalysed oxidative cleavage of 1-octene: yield of heptanoic acid depending on pH-value

Before using "peracetic acid in equilibrium" for the ruthenium-catalysed oxidative $C=C$-cleavage, it is necessary to buffer carefully (0.5–1 mol equivalent Na_2CO_3 related to H_2SO_4), otherwise the yield of carboxylic acid drops remarkable [cf. 51]. The influence of buffering is shown by Fig. 3.

All four oxygen atoms of the carboxylic acid formed by $C=C$-cleavage are derived from the peracetic acid; therefore the molar ratio of olefin(peracetic acid has to be 1/4 or higher. An excess of peracetic acid caused higher yields of carboxylic acid. Compared with "buffered peracetic acid in equilibrium" using distilled peracetic acid as the oxidant results in a higher yield of carboxylic acids. Admittedly to reach these higher yields a greater excess of oxidant was also necessary, because of the faster decomposition of distilled peracetic acid.

Suitable solvents are water and n-hexane; complexing solvents like DMF, acetonitrile and ethers (THF, dioxane) and addition of N-containing complexing agents (EDTA) decreases the yields of carboxylic acids.

Table 6. Ruthenium-catalysed oxidative cleavage of 1-octene — dependence on the molar ratio of olefin to peracetic acid

Ru(acac)$_3$/Olefin/Peracetic acid (molar)	Yield of heptanoic acid (mol-%)	
	A	B
1/1000/4000	70	
1/1000/4400	77	74
1/1000/5000	79	85
1/1000/6000	82	89

A: Buffered peracetic acid in equilibrium. B: Distilled aqueous peracetic acid. Charge: 0.05 mol olefin; 0.05 mmol Ru(acac)$_3$; 20 ml H_2O; 5 h; 65 °C

The influence of various ruthenium compounds is shown in Table 7. Except for Ru-metal and Ru-red all compounds catalyse the oxidative cleavage of 1-octene practically to the same degree. The use of simple ruthenium compounds like RuO_2, $RuCl_3$ or $Ru(acac)_3$ and particular ruthenium complexes and clusters (Fig. 4) clarifies that various ligands (exeption: N-ligands) have no significant influence – at least if a molar ratio of Ru-compound/olefin = 1/1000 is applied.

Table 7. Ruthenium-catalysed oxidative cleavage of 1-octene by peractic acid; various ruthenium compounds as catalysts

catalyst	yields of product heptanoic acid	(mol-%) hexanoic acid	acyloine	epoxide	glycol-mono-acetate
$RuO_2 \cdot nH_2O$	70 (78)	2 (2)	5 (4)	1 (0)	1 (1)
$RuCl_3 \cdot nH_2O$	74 (74)	2 (2)	6 (5)	2 (0)	1 (1)
TPARuO$_4$	68 (68)	2 (1)	5 (3)	1 (0)	4 (0)
$Ru(acac)_3$	66 (77)	2 (2)	6 (6)	3 (0)	1 (2)
$L_{OEt}Ru(CO)_2Cl$	62 (73)	1 (1)	3 (4)	3 (0)	0 (2)
$(L_{OEt})_2Ru(CF_3SO_3)$	(53)		(1)	(4)	(2)
$RuCl_2(PPh_3)_3$	67 (73)	2 (2)	5 (5)		1 (1)
$CpRu(PPh_3)_2Cl$	66 (76)	2 (2)	6 (5)	1 (0)	0 (1)
$(Cp*RuCl_2)_2$	(73)	(2)	(4)	(0)	(2)
Ru-cluster A	74 (77)	2 (2)	8 (7)		0 (2)
Ru-cluster B	74 (74)	3 (1)	8 (8)		0 (1)
$Ru_2(CO)_4(AcO)_2]_n$	73	3	8	1	
$Ru_3(CO)_{12}$	73	2	4	2	
Ru-red	27	2	1	6	
Ru-metal	23	1		13	2
–	3			41	5

Charge: 0.05 mol (5.8 g) 1-octene; 0,05 mmol Ru in catalyst; 0.22 mol buffered peracetic acid; temperature 65 °C; retention time 3 h; solvent: hexane 15 ml (yields in brackets) or water 15 ml. Ru : AcOOH : C=C : 1 : 4400 : 1000. For shape of ruthenium complexes see Fig. 4

The remarkable activity of the ruthenium catalyst for the oxidative C=C-cleavage is shown in Fig. 5. Even if a molar ratio of ruthenium-compound/olefin as low as 1/20000 is chosen, the yield of oenanthic (heptanoic) acid will be 62%.

4.3 Ruthenium-Catalysed Oxidative Cleavage of Terminal-Unsaturated Acids and Their Methylesters

Our "know-how", achieved by the oxidative cleavage of 1-octene, was then used to convert ω-unsaturated fatty acids to dicarboxylic acids.

$Ru(acac)_3$ was the catalyst and distilled peracetic acid or "peracetic acid in equilibrium" was the oxidant. The reaction was carried out in water at 70 °C and

Siegfried Warwel, Michael Sojka and Mark Rüsch gen. Klaas

for 5 h. Educts were free ω-unsaturated acids (9-decenoic, 10-undecenoic, 13-tetradecenoic acid) and their methylesters as well.

$$CH_2=CH-(CH_2)_n-COOR \xrightarrow[+CH_3CO_3H]{[Ru(acac)_3]} HCOOH + HOOC-(CH_2)_n-COOR$$

n = 7, 8, 11   80−90%
R = H, CH₃   isolated yield

By using distilled peracetic acid as oxidant, we were able to isolate the resulting dicarboxylic acids or their methylesters in high yields (Table 8).

If free ω-unsaturated fatty acids were oxidized, the resulting long chain dicarboxylic acids precipitated as white solids after the reaction mixture had been cooled to 0 °C and were isolated by filtration. Their purity, considering the oxidative cleavage of 9 decenoic and 10-undecenoic acid, was about 95%;

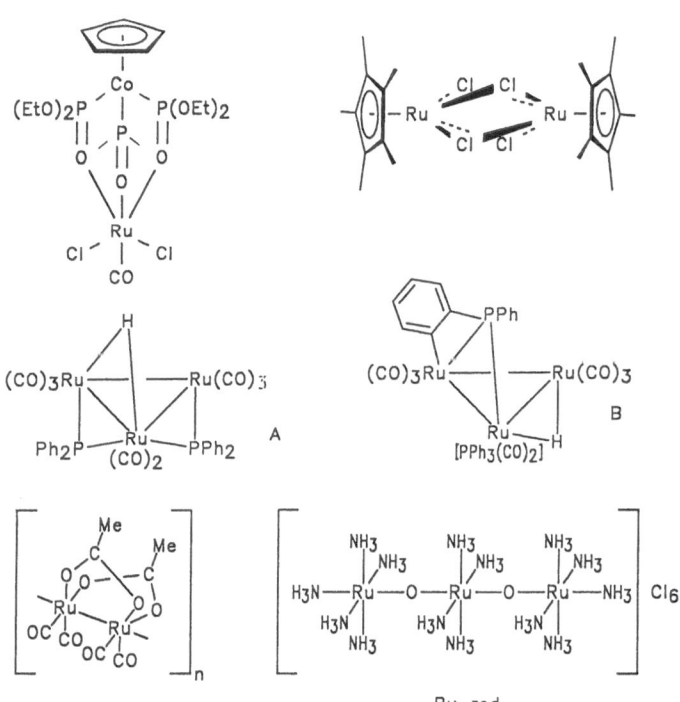

Fig. 4. Various ruthenium compounds tested for their catalytic activity

We would like to thank U. Kölle (Aachen), W. Kläui (Düsseldorf) and G. A. Süß-Fink (Neuchatel, Swiss) for the ruthenium samples.

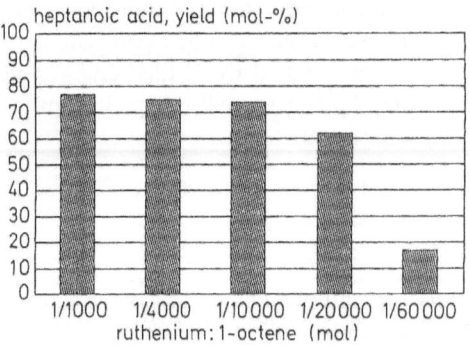

Fig. 5. Ruthenium-catalysed oxidative cleavage of 1-octene by peracetic acid — influence of the catalyst concentration ($Ru(acac)_3$ as catalyst, peracetic acid : olefin = 4.4 : 1 (molar), 65 °C, 3 h)

by-products were shorter dicarboxylic acids. In these oxidations a smaller fraction (about 10 – 15%) of the dicarboxylic acids (acelaic/nonanedioic or sebacic/decanedioic acid) remained in the liquid phase; this fraction was isolated by extraction with ether.

If 13-tetradecenoic acid was cleaved, the whole amount of dicarboxylic acid (brassylic/tridecanedioic acid) precipitated. Its purity was 90% caused by impurities in the educt.

Table 8. Dicarboxylic acids by ruthenium-catalysed oxidative cleavage of terminal-unsaturated fatty acids with distilled peracetic acid

$$CH_2=CH-(CH_2)_n-COOH \xrightarrow[+CH_3CO_3H]{[Ru(acac)_3]} HCOOH + HOOC-(CH_2)_n-COOH$$

n	educt	dicarboxylic acid	yield (%)	purity (%)
7	9-decenoic acid	acelaic acid	83	95
8	10-undecenoic acid	sebacic acid	87	94
11	13-tetradecenoic acid	brassylic acid	88	90

dist. CH_3CO_3H (32%); H_2O as solvent; 70 °C; 3 h
$Ru(acac)_3$: fatty acid : peracetic acid = 1 : 1000 : 5000 (molar)

Table 9. Dicarboxylic monomethylesters by ruthenium-catalysed oxidative cleavage of ω-unsaturated fatty acid methylesters with different types of peracetic acid

Educts	Products	Yield (%)	
		A	B
9-Decenoic acid ME	Nonandioic acid monoME	81	84
10-Undecenoic acid ME	Decandioic acid monoME	83	89
13-Tetradecenoic acid ME	Tridecandioic acid monoME	78	92

ME: Methylester. A: Buffered peracetic acid in equilibrium; yield accordingto GC. B: Distilled aqueous peractic acid (32%); yield, isolated by distillation. Charge: 0.05 mol fatty acid; 0.05 mmol $Ru(acac)_3$; 0.22 mol peracetic acid (A) or 0.25 mol peracetic acid (B) 15 ml H_2O; 5 h; 70 °C; $Ru(acac)_3$/fatty acid/peractic acid = 1/1000/4400–5000 (molar)

Siegfried Warwel, Michael Sojka and Mark Rüsch gen. Klaas

By the same procedure ω-unsaturated fatty acid methylesters can be converted to dicarboxylic acid monomethylesters (Table 9). Using distilled peracetic acid, yields of these bifunctional compounds − now isolated by distillation − were 85−90% and purities were about 94%. Using "peracetic acid in equilibrium" as the oxidant the yields (according to GC) were a little less.

4.4 Catalyst Recycling

In homogeneous transition-metal catalysis, separation and recycling of the catalyst are of the greatest importance. This is especially true for industrial processes involving noble metals.

We were able to achieve such a recycling for the oxidative cleavage of 10-undecenoic acid to the sebacic (decandioic) acid with Ru(acac)$_3$ as catalyst and "peracetic acid in equilibrium". Instead of water or hexane as solvent we applied the high-boiling dodecane for this purpose. After the oxidation had taken place at 70 °C for 5 h, the reaction mixture was cooled to 0 °C and the reaction product (sebaic acid) was isolated by filtration and washed with water/hexane.

All filtrates were taken and hexane, water and acetic acid, formed during the reaction were evaporated, whereas only a solution of the ruthenium catalyst in dodecane remained. This solution was used as the catalyst solution for the oxidative cleavage of 10-undecenoic acid repeatedly. This catalyst recycling was carried out with success seven times (Table 10).

Based on the results a simplified flow diagram for the practical application of this reaction was developed (Fig. 6).

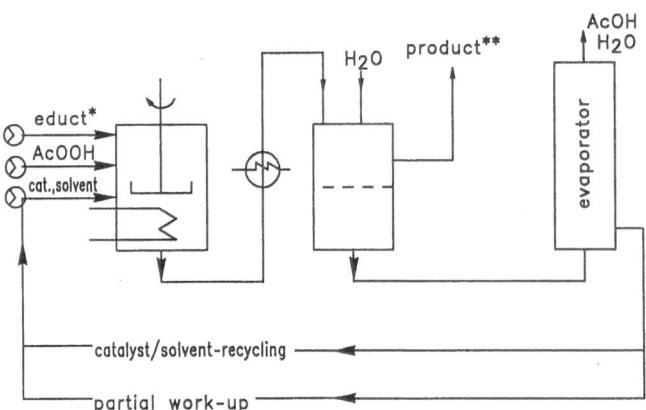

* ω-unsaturated fatty acid   ** α,ω-dicarboxylic acid

Fig. 6. Simplified flow diagram of the ruthenium-catalysed oxidative cleavage of 10-Undecenoic acid with peractic acid and catalyst recycling

94

Table 10. Ruthenium-catalysed oxidative cleavage of 10-undecenoic acid by peractic acid with catalyst recycling

Run	1[a]	2[b]	3[b]	4[b]	5[b]	6[b]	7[b]	8[b]	average
Yield of Decandioic acid (%)	59	81	73	81	75	58	74	80	73

Charge: a) 0.05 mol (9.3 g) 10-undecenoic acid; 0.05 mol (20 mg) $Ru(acac)_3$; 0.22 mol (44 g; 37%) peracetic acid buffered with 0.6 g K_2CO_3; 20 ml dodecane. b) 0.05 mol (9.3 g) 10-undecenoic acid; 0.22 mol (44 g; 37%) peracetic acid buffered with 0.6 g K_2CO_3; 5 ml Dodecane for compensating losses of solvent occuring during the filtration 5 h; 70 °C; $Ru/C = C/peracetic$ acid $= 1/1000/4400$

4.5 On the Way to the Oxidative C=C-Cleavage by H_2O_2 as the Oxidant

In order to apply the oxidative cleavage of ω-unsaturated fatty acids to dicarboxylic acids practically it would be very advantageous with regard to security and economy, if peracetic acid could be replaced by H_2O_2. Such experiments with various ruthenium catalysts and H_2O_2 or acetic acid/H_2O_2 as oxidants were not successful because ruthenium compounds cause a rapid non-productive decomposition of H_2O_2. Therefore we examined the decomposition of H_2O_2 in the presence of several transition metal compounds, various solvents and at different temperatures. Some results are shown in Fig. 7. Whereas $Ru(acac)_3$ caused a rapid decomposition of H_2O_2 even at room temperature, and $Pd(OAc)_2$ decomposes H_2O_2 considerably, H_2O_2 is not decomposed by Re_2O_7 at all. Even at higher temperature (80 °C) H_2O_2 remains stable in the presence of this compound.

Unfortunately the suitability of Re_2O_7 as an oxidation catalyst is scarcely known. G. W. Parshall succeded in the oxidative cleavage of cyclododecene by H_2O_2 in acetic acid, but the yield of the product (dodecanedioic acid) was only

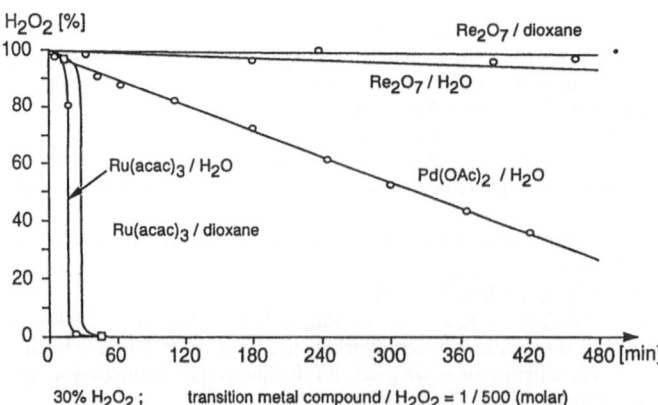

Fig. 7. Decomposition of H_2O_2 by Re_2O_7, $Pd(OAc)_2$ and $Ru(acac)_3$

Siegfried Warwel, Michael Sojka and Mark Rüsch gen. Klaas

about 30% [54]. CH_3ReO_3, well accessible by the reaction of Re_2O_7 and $Sn(CH_3)_4$ [15], displays a totally different behaviour; as W. A. Herrmann et al. were able to show, it is an excellent catalyst for epoxidation and hydroxylation of olefines in H_2O_2/*tert*-butyl alcohol [17, 55, 56]. But also in this context, Re_2O_7 itself is described as ineffective [55].

Recently we found that Re_2O_7 catalysed the hydroxylation of olefins by H_2O_2, if appropriate solvents like 1,4-dioxane or trialkylphosphate are applied and if the reaction takes place at higher temperatures [57]:

$$R^1-CH=CH-R^2 \ + \ H_2O_2 \ \xrightarrow[\substack{90\ °C,\ 1,4\text{-dioxane}}]{[Re_2O_7]} \ R^1 \ \underset{OH}{\overset{OH}{\underset{|}{\overset{|}{CH}}}} \ CH \ R^2$$

Applying a molar ratio of Re_2O_7/olefin/H_2O_2 (60%) as 1/100/120, the diols were isolated in yields of 60–80%. An easy catalyst recycling is therefore shown to be possible and advantageous.

Encouraged by these results further research on the application of Re_2O_7 as an oxidation catalyst is in progress.

Acknowledgement. We gratefully acknowledge support of this work from the Federal Minister of Research and Technology of Germany (BMFT) and the Peroxid-Chemie GmbH, Höllriegelskreuth.

5 References

1. Baumann H, Buhler M, Fochem H, Hirsinger F, Zoebelein H, Falbe J (1988) Angew Chem 100: 41
2. Harries C (1905) Liebigs Ann Chem 343: 311
3. Criegee R (1975) Angew Chem 87: 765
4. Heins A, Witthaus M (1984) Henkel-Referate 20: 42
5. Zaidman B, Kisilev A, Sasson Y, Garti N (1988) J Amer Oil Chem Soc 65: 611
6. Banks RL, Bailey GC (1964) Ind Eng Chem, Prod Res Develop 3: 170
7. a) Ivin KJ (1983) Olefin metathesis, Academic, London; b) Dragutan V, Balaban AT, Dimonie M (1985) Olefin metathesis and ring-opening polymerization of cyclo-olefins, J Wiley, Chichester, Editura Academiei, Bukarest
8. Warwel S (1987) Erdöl − Erdgas − Kohle, 103: 238
9. van Dam PB, Mittelmeijer MC, Boelhouwer C, J Chem Soc, Chem Commun 1972: 1221
10. Reviews: a) Boelhouwer C, Mol JC (1985) Prog Lipid Res 24: 243; b) Mol JC (1991) J Mol Catal 65: 145
11. Verkuijlen E, Kapteijn F, Mol JC, Boelhouwer C, J Chem Soc, Chem Commun 1977: 198
12. Warwel S, Jägers H-G, Deckers A (2. 3. 90/5. 9. 91) Ger Offen DE4006540
13. Warwel S (1992) Nachr Chem Tech Lab 40: 314
14. Warwel S, Deckers A, Döring N, Ercklentz B, Jägers H-G, Thomas S (1992) In: Statusbericht zum BMFT-Forschungsverbundvorhaben „Neue Einsatzmöglichkeiten nativer Öle und Fette als Chemierohstoffe". BMFT-Projektträger Biologie-Energie-Ökologie, Forschungszentrum Jülich GmbH, Jülich, p 83
15. Herrmann WA, Kuchler JG, Felixberger JK, Herdtweck E, Wagner W (1988) Angew Chemie 100: 420

16. Herrmann WA, Wagner, W Volkhardt U (5. 12. 1989/13. 6. 1991) (Hoechst AG) Ger Offen DE3940196
17. Herrmann WA (1990) J Organometal Chem 382: 1
18. Herrmann WA, Wagner W, Flessner UN, Volkhardt U, Komber H (1991) Angew Chem 103: 1704
19. Warwel S, Deckers A, Jägers H-G, Ercklentz B, Harperscheid M, Angew Chem, in preparation
20. Warwel S, Pompetzki W, Deckwirth EA (1991) Fat Sci Technol 93: 210
21. Walens HA, Koob RP, Ault WC, Maerker G (1965) J Amer Oil Chemists Soc 42: 126
22. Smidt J, Hafner W, Jira R, Sedlmeier J, Sieber R, Rüttinger R, Kojer H (1959) Angew Chem 71: 176
23. Herrmann WA Kontakte (Darmstadt) 1991: (1) 22
24. Clement WH, Selwitz CM (1964) J Org Chem 29: 241
25. Dzhemileva GA, Odinokow NV, Dzhemileva UM, Tolstikov GA (1983) Bull Acad Sci USSR 32: 307
26. Subramaniam CS Synthesis 1987: 468
27. McQuillin FJ, Parker DG, J Chem Soc, Perkin I 1975: 2092
28. Weissermel K, Arpe HJ (1988) Industrielle organische Chemie, 3rd edn, VCH, Weinheim, p 256
29. Steadman ThR, Peterson JOH (1959) (National Research Corp), US Pat 2847432 (12. 8. 58), CA 53: 4143
30. O'Brien DA (1986) (Procter & Gamble Comp), EP 153522 (4. 9. 85), CA 104: 131950
31. Djerassi C, Engle R (1953) J Amer Chem Soc 75: 3838
32. Berkowitz LM, Rylander PN (1958) J Amer Chem Soc 80: 6682
33. Lee DG (1969) In: Augustine RL (ed) Oxidation, vol I, Marcel Dekker, New York
34. Sheldon RA, Kochi JK (1981) Metal-catalyzed oxidations of organic compounds, Academic, New York pp 162, 297
35. Gore ES (1983) Platinium metals Rev 27: 111
36. Seddon EA, Seddon KR (1984) In: Clark RJH (ed) The chemistry of ruthenium. Collection of monographs no 19, Elsevier, Amsterdam, p 52
37. Haines AH (1985) Methods for the oxidation of organic compounds, Academic Press, London p 128
38. Courtney JL (1986) In: Mijs WJ, de Jonge CRHI (ed) Organic syntheses by oxidation with metal compounds, Plenum Press, New York p 445
39. Hudlicky M (1990) Oxidations in organic chemistry, ACS Monograph 186, American Chemical Society, Washington p 82
40. Carlsen PHJ, Katsuki T, Martin VS, Sharpless KB (1981) J Org Chem 46: 3936
41. Starks ChM, Washecheck PH (1971) (Continental Oil Co) US Pat 3547962 (19. 12. 1968/ 15. 2. 1970), CA 74: 140895
42. Kebyls KA, Dubeck M (1969) (Ethyl Corp), US Pat 3409649 (14. 12. 1964/5. 11. 1968), CA 70: 114575
43. Wolfe S, Hasan SK, Campball JR, J Chem Soc, Chem Commun 1970: 1420
44. Lee DG, Chang VS, Helliwel S (1976) J Org Chem 41: 3644
45. Foglia TA, Barr PA, Malloy AJ, Costanzo MJ (1977) J Am Oil Chem Soc 54: 870
46. Foglia TA, Barr PA, Malloy AJ (1977) J Am Oil Chem Soc 54: 858
47. Mac Lean AF, Stautzenberger AL (1969) (Celanese Corp) Ger Offen DE 1568363 (29. 12. 1966/9. 7. 1970), CA 70: 57090
48. Mac Lean AF (1970) (Celanese Corp) Ger Offen DE 1568364 (30. 12. 1966/2. 4. 1970), CA 72: 21323
49. Washecheck PH (1971) (Continental Oil Co), Ger Offen DE 2046034 (17. 9. 1970/19. 5. 1971), CA 75: 35158
50. Merk W, Schreyer G, Weigert W (1972) (Degussa AG) Ger Offen DE 2106307 (10. 2. 1971/31. 8. 1972), CA 77: 151485
51. Sheng MN (1974) (Atlantic Richfield Co) US Pat 3839375 (9. 11. 1972/1. 10. 1974), CA 81: 269145

52. a) Weiert WM, Merk W, Offermanns H, Prescher G, Schreyer G, Weiberg O (1975) Chemiker-Ztg 99: 111; b) Weigert WM, Merk W, Offermanns H, Prescher G, Schreyer G, Weiberg O (1978) In: Weigert WM (ed) Wasserstoff-Peroxid und seine Derivate, Hüthig, Heidelberg, p 73
53. "Proxitane" and "Proxilate", Product information of Peroxid Chemie GmbH, Höllriegelskreuth, Germany
54. Parshall GW (1972) (du Pont), US Pat 3646130 (15. 5. 70/29. 2. 72), CA 76: 154392
55. Herrmann WA, Marz D, Wagner W, Kuchler JG, Weichselbaumer G, Fischer R, Ger Offen DE 3902357 (27. 1. 89/2. 8. 90), Hoechst AG
56. Herrmann WA, Fischer RW, Marz DW (1991) Angew Chem 103: 1706
57. Warwel S, Rüsch gen Klaas M, Sojka M, J Chem Soc, Chem Commun 1991: 1578

Peracids and Free Radicals:
A Theoretical and Experimental Approach

Jacques Fossey[1], Daniel Lefort[2] and Janine Sorba[1]

[1] Laboratoire des Mécanismes Réactionnels, URA 1307, Ecole Polytechnique, DCMR, 91128 Palaiseau, France
[2] LECSO; UMR 28 CNRS, BP 28, 94320 Thiais, France

Table of Contents

Organic peracids behave as a source of alkyl radicals and serve as a good model for studying their chemistry. The OH transfer by alkyl radicals and the stereoelectronic control of the reaction were investigated in detail. Inter- and the intramolecular hydrogen shifts by alkyl radicals were also studied. These results were used in order to undertake selective hydroxylation of hydrocarbons. Finally we turned our attention to the reactivity of electrophilic radicals towards peracids.

Topics in Current Chemistry, Vol. 164
© Springer-Verlag Berlin Heidelberg 1993

1 Introduction

Organic peracids are mainly used in oxidation reactions such as epoxidation, Baeyer-Villiger reactions etc. Some years ago [1–30] we discovered that peracids in hydrocarbons at reflux temperature undergo thermal decomposition mainly into alcohol (Table 1 − runs 1 and 2). If the reaction takes place at just below reflux temperature (Table 1 − runs 3 and 4), the product distribution depends on the gaseous atmosphere: under argon, peracids decompose mainly into alcohol (Table 1 − run 3) [31] but under air they slowly give quantitatively the corresponding acids (Table 1 − run 4) [1, 20]. At low temperatures (55 °C), in the presence of an initiator such as di(4-tert-butylcyclohexyl)peroxydicarbonate (BCHPC) and under argon, the reaction leads to good yields of alcohol (Table 1 − runs 6 and 7) [31].

Table 1. Decomposition of peracids RCH_2CO_3H, 0.1 M (RCH_2 = decanyl-1 or nonyl-1)

run	solvent	gaseous atmos-phere	T °C	$t_{1/2}$ min	product concentrations in moles per 100 moles of peracid							
					RCH_2CO_2H	RCH_2OH	RCH_3	$R^{ih}OH$	RCO_2H $+RCH=O$	$R(-H)$	$CyOH^a$ $+Cy=O^a$	PhR
1	CyH[a]	air	100	1	9	53	23	8	4	0	21	
2	PhH	air	100	7	40	27	2	4	9	0		2
3	CyH[a]	argon	78	35	28	53	10	4	8	1	17	
4	CyH[a]	air	78	840	93	1	1	0	3	3	26	
5	CyH[a]	argon	55[b]	6	0	58	19	0	0	2	30	
6	PhH	argon	55[b]	20	13	65	7	0	5	7		

[a] CyH = cyclohexane. CyOH = cyclohexanol. Cy = O cyclohexanone
[b] In presence of di(4-tert-butyl cyclohexyl)peroxydicarbonate (BCHPC), 0.1 M

2 Mechanism Leading to the Alcohol ROH

Kinetic analysis [1, 3–5, 7, 9], use of inhibitors [1–2, 4–5, 7, 9], CIDNP experiments [16, 19] and the stereochemistry [8, 12–13, 17–18, 30] indicate that the degradation of peracids RCO_3H into alcohols ROH takes place through a free-radical chain mechanism (cf. Schemes 1 and 2). The initiation step (reaction 1) is the thermal homolysis of the weak $-O-O-$ bond of the peracid [32]. The two propagating steps forming the alcohol are:
− decarboxylation of $RCO_2^{\cdot}$ into $R^{\cdot}$ and CO_2 (reaction 2),
− an S_H2 reaction of $R^{\cdot}$ upon the peroxidic oxygen (reaction 3).
  Then the characteristic chain termination reaction occurs (reaction 4). Usually the chain length is over 1000.
  Below reflux temperature and under air (Table 1 − run 4), O_2 rapidly traps the alkyl radicals $R^{\cdot}$ ($k > 10^9$ M^{-1} s^{-1}) [33] and stops the chain mechanism [20]. The reaction then slows down and peracids are transformed into their corresponding acids by mechanisms described later.

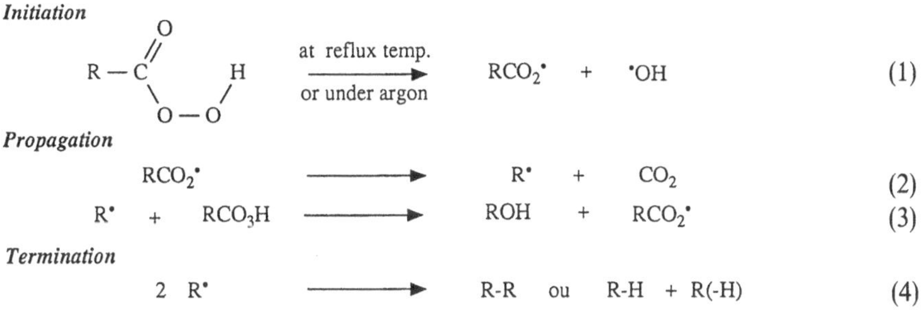

Initiation

$$R-C\underset{O-O}{\overset{O}{<}}H \xrightarrow[\text{or under argon}]{\text{at reflux temp.}} RCO_2^{\cdot} + {\cdot}OH \tag{1}$$

Propagation

$$RCO_2^{\cdot} \longrightarrow R^{\cdot} + CO_2 \tag{2}$$

$$R^{\cdot} + RCO_3H \longrightarrow ROH + RCO_2^{\cdot} \tag{3}$$

Termination

$$2\ R^{\cdot} \longrightarrow R\text{-}R \quad \text{ou} \quad R\text{-}H + R(\text{-}H) \tag{4}$$

Scheme 1. Chain mechanism for ROH formation

The degradation of the peracid into alcohol can be very efficient. So adamantane peracid in the bridgehead position decomposes rapidly into the corresponding alcohol in cyclohexane as well as in benzene with yields higher than 90% [21, 29] (reaction (5)).

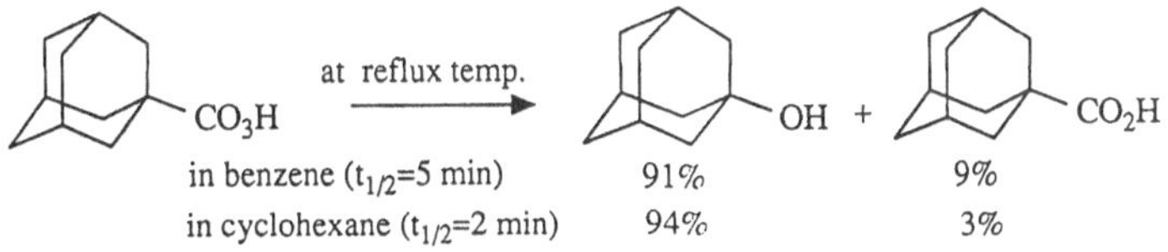

$$\tag{5}$$

in benzene ($t_{1/2}$=5 min) 91% 9%
in cyclohexane ($t_{1/2}$=2 min) 94% 3%

Yields of alcohol depend strongly on the nature of the peracids [21, 24, 26, 28–30]. Good results are obtained with tertiary peracid [21] but the bicyclo[2.2.1]heptyl-1-peroxycarboxylic acid gives a low quantity of alcohol [14, 21]. Notice that perbenzoic acid does not decompose into phenol [5, 20, 29].

Table 2. Peracid decomposition products (0.1 M in cyclohexane at reflux). $R^1 = CH_3(CH_2)_9$. Results in mole per 100 moles of peracid

R =	$R^1\overset{\cdot}{C}(CH_3)_2$	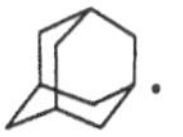 (adamantyl)$^{\cdot}$	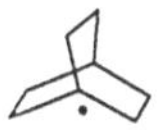 (bicyclo)$^{\cdot}$	 (bicyclo)$^{\cdot}$	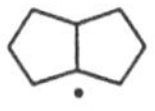 (bicyclo)$^{\cdot}$	Ph$^{\cdot}$
ROH	89	94	87	17	95	0
RH	0	3	12	79	1	40
RCO$_2$H	11	3	1	5	4	60
CyOH^a	1	4	15	77	2	95

	$R^1\overset{\cdot}{C}HCH_3$	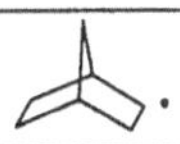 (bicyclo)$^{\cdot}$	 (alkenyl)$^{\cdot}$	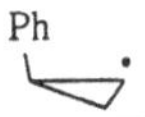 Ph (cyclobutyl)$^{\cdot}$	$R^1\overset{\cdot}{C}H_2$	$Ph\overset{\cdot}{C}H_2$
ROH	97	91	97	0	54	27
RH	1	3	2	79	23	0
RCO$_2$H	-	3	1	22	9	52
CyOH^a	3	3	2	75	21	0

^{a} CyOH = cyclohexanol

In a free-radical chain mechanism we need three types of component. Here the peracid plays the following three roles:
— initiator in reaction (1),
— source of radical R° in reaction (2),
— substrate in reaction (3).
Once the peracid is prepared, this reaction is very easy to handle.

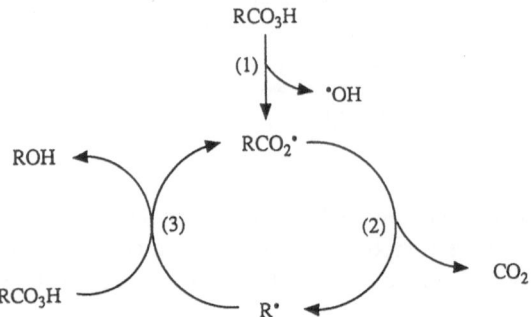

Scheme 2. Cyclic diagram for ROH formation in free-radical peracid decomposition

3 An S_H2 Reaction of Alkyl Radicals: The OH Transfer from the Peracid

The interesting point of this chain mechanism is step (3), scheme 2, where an alkyl radical R° reacts with the peroxidic oxygen of the peracid in order to transfer the OH group and form an alcohol ROH.

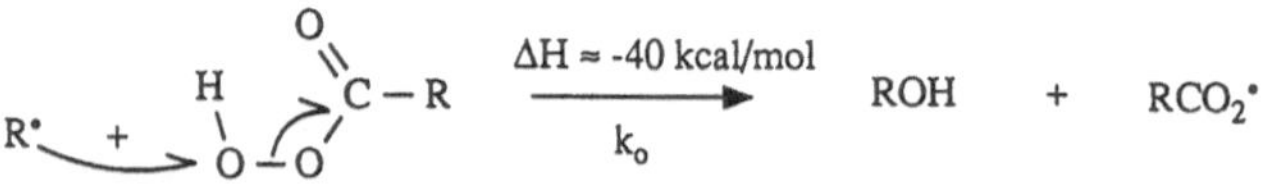

$$R° \;+\; \underset{O-O}{\overset{H}{\underset{|}{\diagup}}}\!\!C-R \quad\xrightarrow[k_0]{\Delta H \approx -40\ \text{kcal/mol}}\quad ROH \;+\; RCO_2° \tag{3}$$

This reaction is so fast that it was a longtime before we succeeded for the first time in trapping the nopinyl radical before its opening [30] (cf. scheme 3).

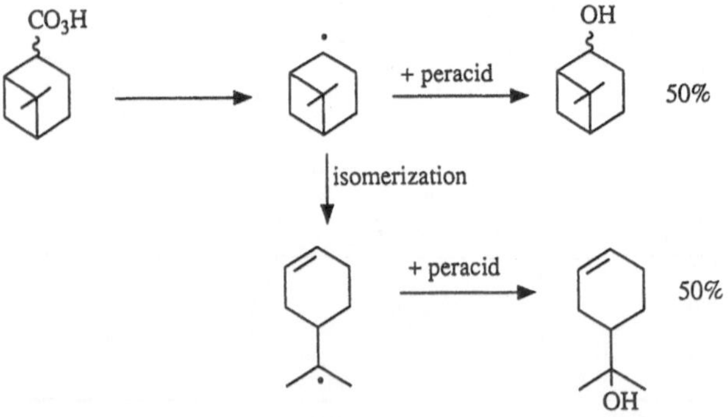

$$(6)$$

isomerization

$$(7)$$

Scheme 3. Alcohol formation in nopinane-1-peracid decomposition

From theoretical calculations [30], the orbital bearing the single electron of the radical is coming along the $O-O$ axis in order to get the better overlap with the σ and σ^* orbitals describing the $O-O$ peroxidic bond. So, the $C-O$ bond formation of the alcohol accurs simultaneously with the breaking of the $O-O$ bond (S_H2 reaction).

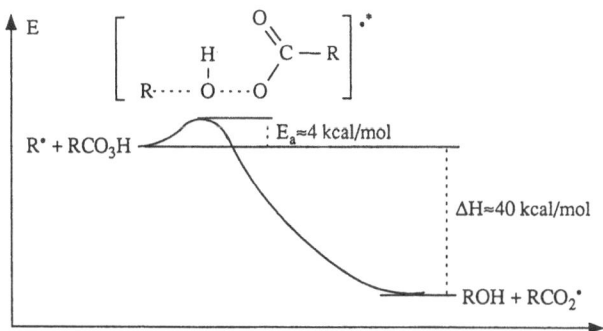

Scheme 4. Energy profile for reaction (3) $R^{\cdot} + RCO_3H$

Due to the weakness of the $O-O$ bond (≈ 47 kcal/mol) [32] the exothermicity of reaction (3) is important (≈ 40 kcal/mol). For the bicyclo[2.2.1]heptyl-1 radical the activation energy E_a was estimated experimentally to be around 4 kcal/mol [21]. As this radical is one of the less reactive ones, E_a is generally lower for most of the radicals. We find a good correlation between the Ionization Potential of $R^{\cdot}$ ($IP_{R^{\cdot}}$) and the rate constant k_0 for reaction (3) [21] (Scheme 5).

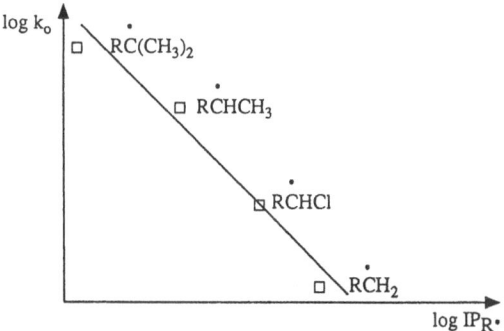

Scheme 5. Correlation between $\log IP_{R^{\cdot}}$ and $\log k_0$ for reaction (3)

It should be emphasized that reaction (3) is not governed by the radical stability since the more stable radicals are the more reactive ones. So a tertiary alkyl radical reacts faster than a primary one. The correlation shown in Scheme 5 indicates that reaction (3) is controlled by the energy gap between the radical SOMO and

LUMO of the $O-O$ bond i.e. the energy of the low lying σ^*_{O-O} orbital (Scheme 6). The energy effect associated with the SOMO-LUMO interaction becomes more favorable as the energy gap decreases. So tertiary radicals having a higher SOMO than secondary and primary radicals are more reactive. If the reaction were controlled by an orbital interaction between the SOMO and an occupied MO of the peracid group, i.e. the σ_{O-O} MO, the reverse reactivity should have been observed; the primary radical should have been more reactive than the tertiary one.

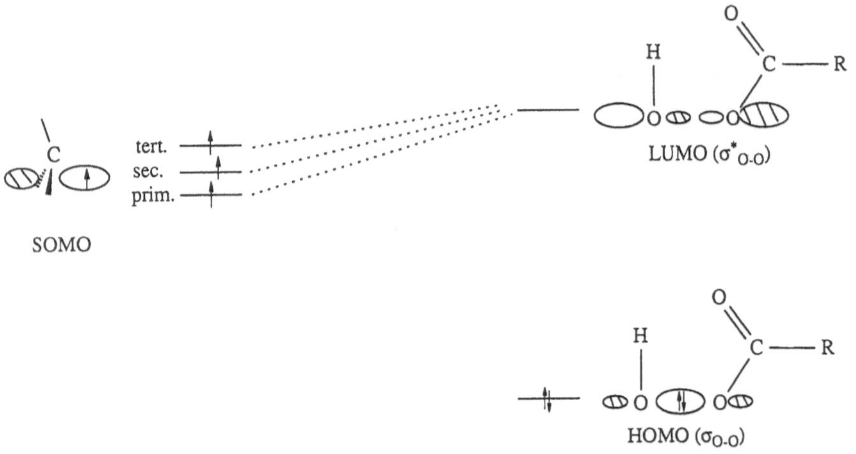

Scheme 6. Interaction diagram between frontier orbital for reaction (3).

An alternative explanation is given by the classical free-radical *polar effect* which figures out the transition state of reaction (3) by two limit structures. The second one implies an electron transfer from the radical SOMO into the peracid $O-O$ LUMO.

Scheme 7. Polar effect in reaction (3) transition state.

Consequently, reaction (3) should depend on the stability of the cation corresponding to the radical R˙. This is clearly the case, as shown in Scheme 8 by the good correlation obtained between the stability of the cation (measured by the S_N1 rate constant) and k_0 [21]. This interpretation explains why radicals having a σ configuration (pyramidalized radicals) such as the cyclopropyl or the bicyclo[2.2.1]heptyl-1 [34], less nucleophilic than π radicals (quasi planar radicals having their unpaired electron in a p orbital), give no or only small quantities of the corresponding alcohols [24, 29] (Table 2). On the other hand, benzyl radicals (Table 2), having a lower I.P. than a primary alkyl radical should be more nucleophilic. Actually, due to the delocalization of the single electron upon the aromatic ring, this radical reacts slowly with the peracid [29].

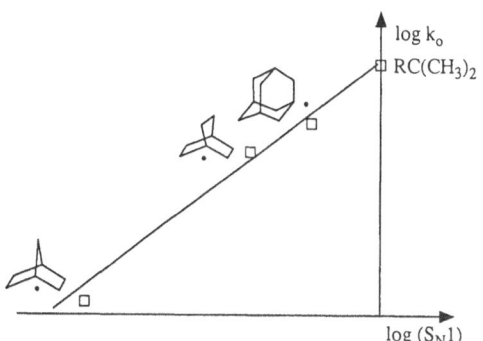

Scheme 8. Correlation between $\log k(S_N1)$ and $\log k_0$ for reaction (3)

4 Stereoelectronic Control for OH Transfer — Reaction (3)

An Attack of the alkyl radical upon the peracid can be under stereoelectronic control since the OH group can enter by one of the two faces of the alkyl radical (Scheme 9).

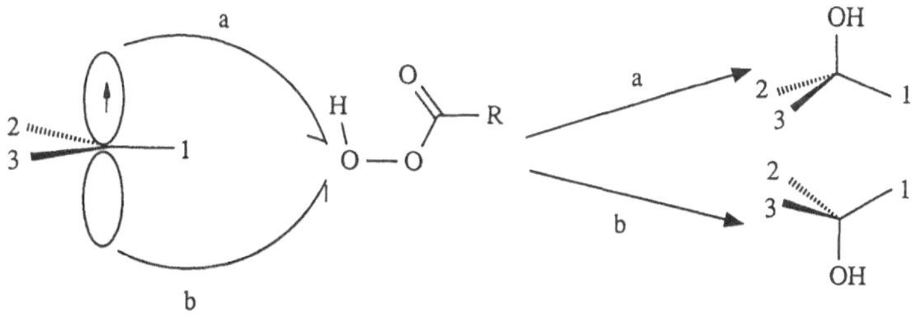

Scheme 9. Stereocontrol in a radical attack upon peracid

Some examples [8, 12–13, 17–18, 30] of stereoelectronic control are given in Scheme 10.

The stereocontrol with the 4-*tert*-butylcyclohexyl radical is well explained by torsional effects [35–36] which are unfavorable for the OH entry be the equatorial face (Scheme 11) [12]. This due to the small dihedral angle between the $C_1 - OH$ forming bond and the $C_2 - H_2ax.$ and $C_6 - H_6ax.$ bonds. On the other hand, torsional effects on the axial face are weak. When $R_3 = Me$ the steric hindrance

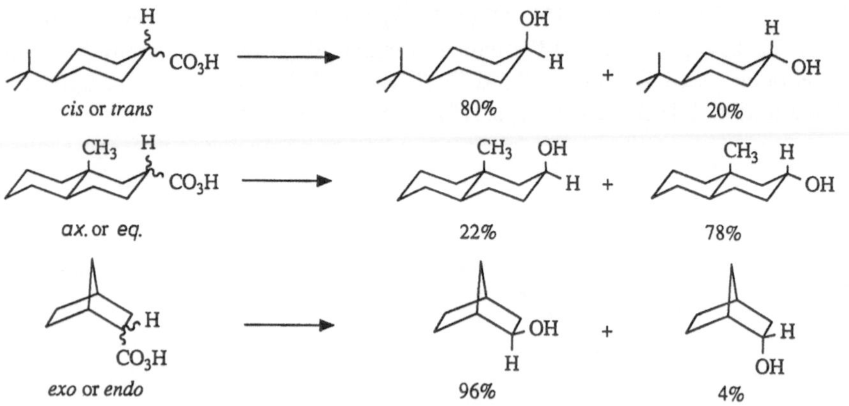

Scheme 10. Stereocontrol in radical attacks on the peracid

is increased on the axial face and a reverse stereocontrol is observed. When R_6ax. = Me the torsional effects, increased on the equatorial face, lead to a better stereocontrol. When R_6eq. = Me we have more hindrance in the nodal plan of the radical SOMO which has little effect on the stereoselectivity.

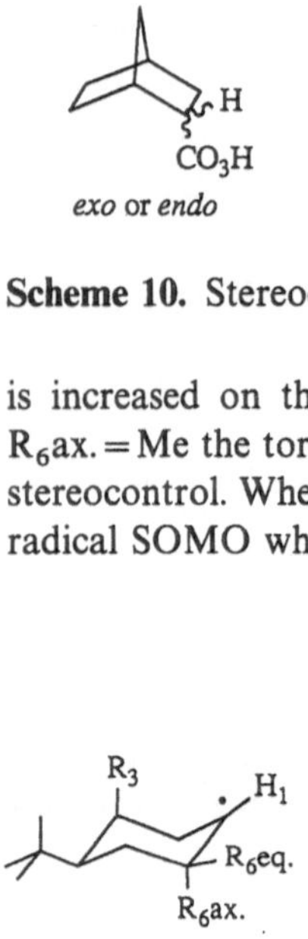

R_3	R_6ax.	R_6eq.	ax/eq
H	H	H	80/20
Me	H	H	25/75
H	Me	H	90/10
H	H	Me	75/25

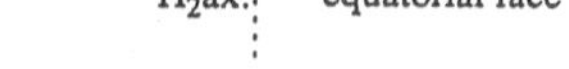

Scheme 11. Stereoselectivity in 4-*tert*-Butylcyclohexyl attack peracid

With the norbornyl-2 radical, the high stereoselectivity in favor of the *exo* alcohol [19] is well explained by the free-radical structure. This one, clearly pyramidal [37–38], with the expansion of the orbital towards the *exo* direction (Scheme 12).

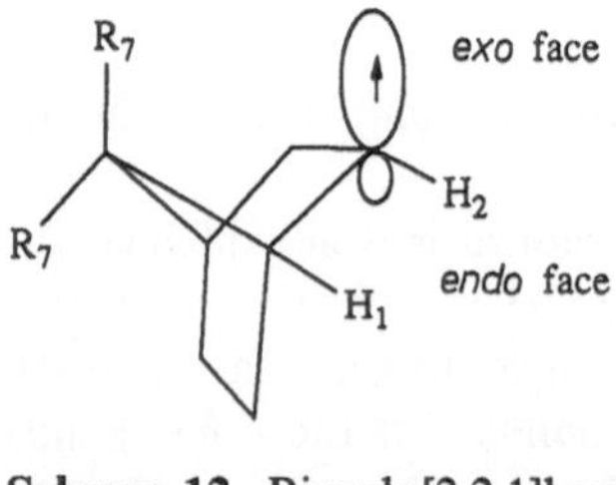

Scheme 12. Bicyclo[2.2.1]heptyle-2 radical structure.

In the nopinyl radical the α face is at a clear disadvantage due to the presence of the methyl group (Scheme 13) [30]. We should notice that the stereoselectivity is the reverse of that obtained by the reduction of the ketone and the less stable β-nopinol is the major reaction product [39]. Therefore the reaction is under kinetic control which is in agreement with the exothermicity of the reaction.

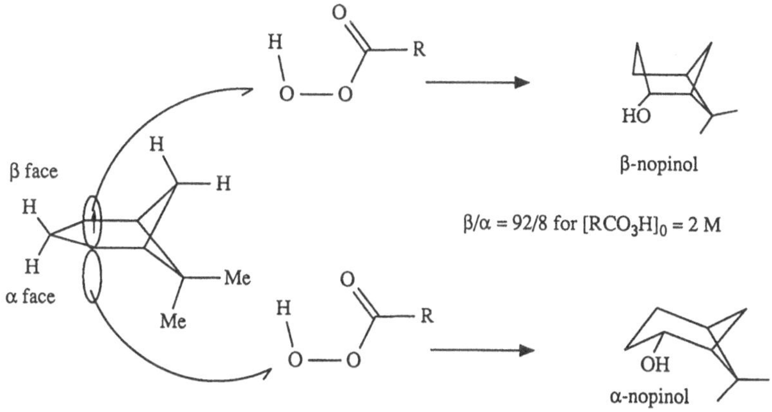

β-nopinol

β/α = 92/8 for $[RCO_3H]_0 = 2$ M

α-nopinol

Scheme 13. Nopinyl radical stereoselectivity

5 Hydroxylation of Hydrocarbons

Up to now we have focused our attention on the alcohol formation by reaction (3) and the characteristics of that reaction. Nevertheless alcohol ROH is not always the main product. For example [14, 21, 29] the bicyclo[2.2.1]heptyl-1-peracid (reaction (8)) gives only acid when it is decomposed in benzene at the reflux temperature. In cyclohexane at the reflux temperature it leads mainly to norbornane and cyclohexanol. Furthermore the reaction is slow in benzene and fast in cyclohexane.

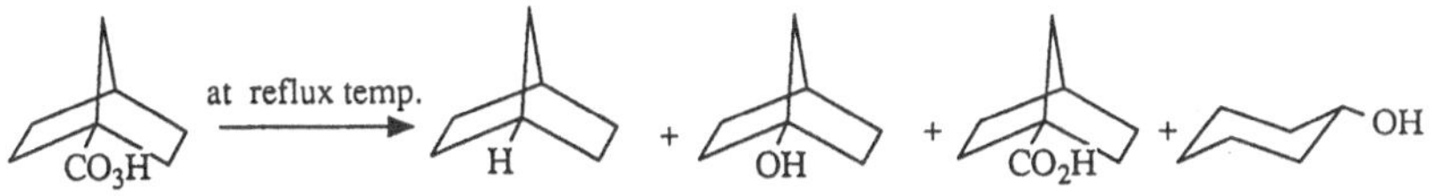

		H	OH	CO_2H	OH
in benzene ($t_{1/2}$=40 min)		4%	0%	98%	
in cyclohexane ($t_{1/2}$= 8 min)		79%	17%	5%	77%

$$(8)$$

In cyclohexane, formation of norbornane, and R−H from a general point of view, is explained by the chain mechanism described on Scheme 14.

$RCO_2^•$ undergoes a fast decarboxylation (reaction (2)) [40]. R⁺ can react either according to reaction (3) or transfer H from the cyclohexane (reaction (9)). The competition between the two reactions depends on the structure of R⁺. Due to the low nucleophilic character of the norbonyl-1 radical, reaction (3) is slow (vide

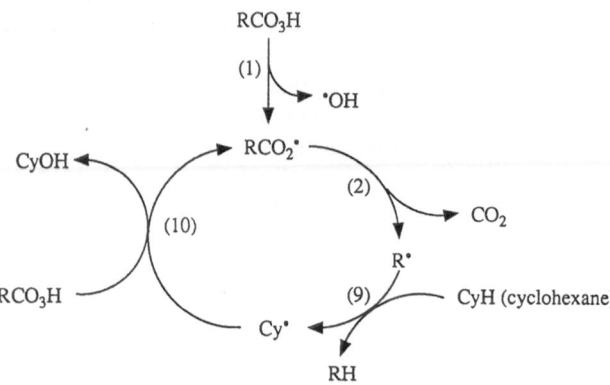

Scheme 14. Cyclic diagram for RH and CyOH formation in free-radical peracid decomposition

supra) and H abstraction of cyclohexane which is an exothermic reaction occurs preferentially. Then the nucleophilic cyclohexyl radical, Cy˙, reacts easily with peracid to give cyclohexanol. Consequently CyOH and RH are formed in equimolecular concentrations (cf. Table 2).

Perbenzoic acid behaves similarly [29]. In benzene at the reflux temperature we obtain, slowly and quantitatively, benzoic acid. In cyclohexane at the reflux temperature, the reaction is faster and, besides benzoic acid, we get benzene and cyclohexanol [20]. The final concentration in CyOH is almost equal to the initial concentration in $PhCO_3H$.

$$Ph-CO_3H \xrightarrow{\text{at reflux temp.}} Ph-CO_2H + Ph-H + cyC_6H_{11}-OH$$

$$\text{in benzene } (t_{1/2} = 20 \text{ min}) \quad 98\% \tag{11}$$

$$\text{in cyclohexane } (t_{1/2} = 2 \text{ min}) \quad 60\% \qquad 40\% \qquad 95\%$$

The chain mechanism which describes the perbenzoic decomposition in cyclohexane at the reflux temperature, becomes more complex but easy to understand [20, 29]. First of all, as $PhCO_2^˙$ slowly loses CO_2 [41], it can lead (before decarboxylation), by H abstraction from the solvent, to $PhCO_2H$ and the cyclohexyl radical (reaction (14)). Ph˙ formed by reaction (12), has a low nucleophilic character [34]. Consequently it cannot transfer OH from the peracid but abstracts H from cyclohexane leading to PhH. Finally, Cy˙ obtained by reactions (12)–(13) or (14) gives CyOH via reaction (15).

We used this type of reaction in order to undertake hydroxylation of hydrocarbon [26]. Under the following conditions, perbenzoic (0.1 M), hydrocarbon (10 M) at the reflux temperature in benzene, we obtain good yields of hydroxylation. Tertiary alcohols are obtained with better yields than secondary or primary ones because abstraction of tertiary H is easier as also is the reaction of the tertiary radicals on the peracid which is controlled by the nucleophilic character of the radical (vide supra). For the *trans* and *cis* decalin, the tertiary alcohols are obtained with retention of the configuration [27].

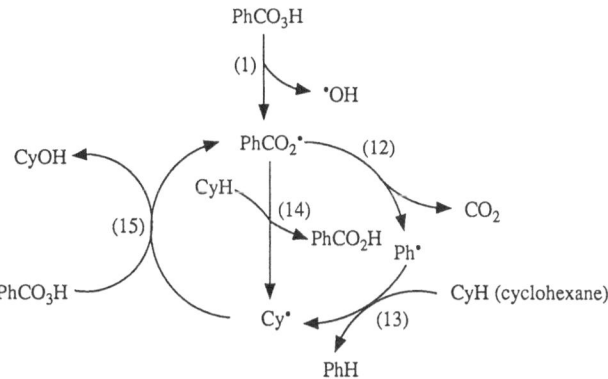

Scheme 15. Cyclic diagram for free-radical perbenzoic decomposition in cyclo-hexane

Table 3. Alcane hydroxylation. Product concentration (ROH) in moles per 100 moles of peracid

PhCO$_3$H	+	RH	at the reflux temperature in benzene	PhCO$_2$H	+	ROH	(16)
0.1 M		10 M					

RH	![cyclohexane H H]	![methylcyclohexane CH$_3$ H]	![adamantane H]	![decalin H]	![H ... H]
ROH	22	37	60	65	56

6 Hydrogen Shift in Alkyl Radicals

With long chain peracids, at low concentration and high temperature, besides ROH, RH and RCO$_2$H, a certain quantity of isomerized alcohols was observed [11, 15, 23, 25] (Scheme 16). The isomerization is regioselective and, from perdodecanoic acid gives only 5-undecanol and 6-undecanol in a ratio of 3.3. The ratio is temperature independent but the concentration of secondary alcohols over primary alcohols increases with the temperature (from 0.23 in pentane to 1.7 in octane in both cases at the reflux temperature with an initial peracid concentration of 0.01 M) [11, 15]. This result is well explained by a regioselective intramolecular hydrogen transfer occuring through a cyclic transistion structure with a CHC angle of around 150° [42] (Scheme 17). This reaction takes place because primary radicals react slowly with peracid. Nevertheless as the activation energy is higher for the isomerization reaction than for the OH transfer from the peracid, low concentration of peracid and higher temperature favored, of course, the iso-merization.

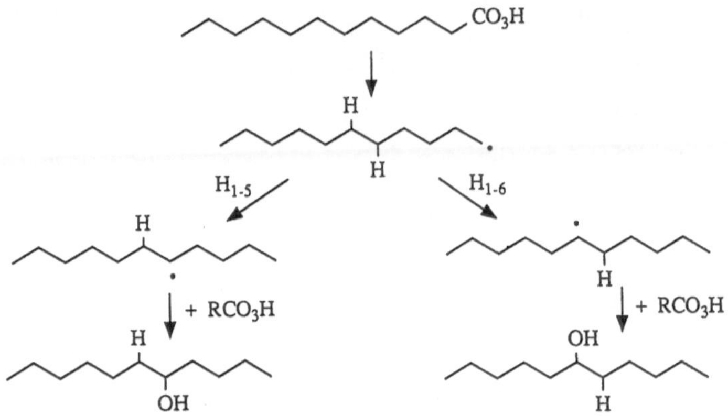

Scheme 16. Formation of secondary alcohols

1,6 H shift   1,5 H shift

Scheme 17. Transition structures for 1,6 and 1,5 H transfer

We used these conditions (low concentration and high temperature) in order to obtain a remote hydroxylation of a steroid D-ring (cf. Scheme 18). Thus the synthesis of 16-hydroxy norcholane in 35% yield is the result of a regioselective intramolecular 1,5-H shift [23].

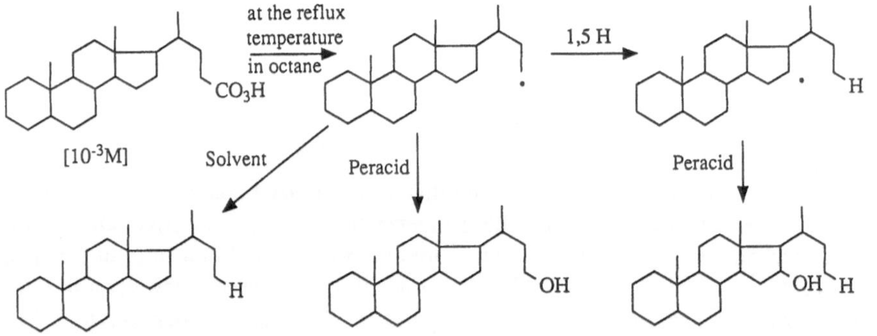

Scheme 18. Peroxycholanic acid decomposition products

7 Free-Radical Decomposition to the Acid

In benzene at the reflux temperature the decomposition of bicyclo[2,2,1]heptyl-1 peracid or benzoic peracid leads slowly to the corresponding acids. In order to explain the acid formation we have two alternative mechanisms. Both imply that the HO˙ radical reacts either by addition on to the C=O of the peracid group (Scheme 19) or by abstraction of the peroxydic H (Scheme 20) [20].

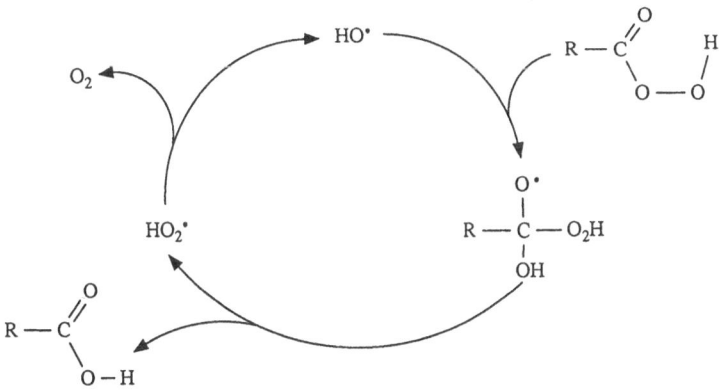

Scheme 19. Acid formation via HO˙ addition onto the C=O of the peracid

After the addition, the tetrahedric radical leads, by a β-fragmentation, to the acid and $HO_2^˙$. Two $HO_2^˙$ radicals dimerize, then fragment into O_2 and HO˙.

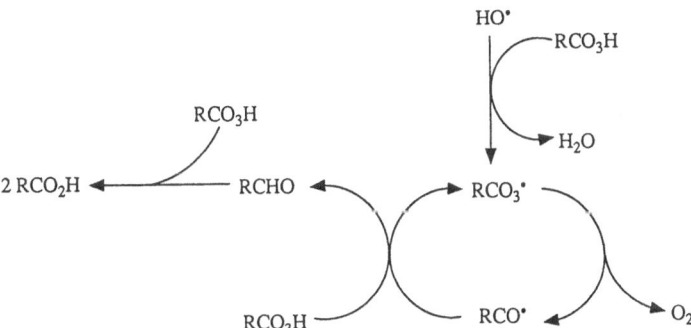

Scheme 20. Acid formation via H abstraction from the peracid

In Scheme 20, HO˙ by H-abstraction from the peracid leads to the acylperoxide radical $RCO_3^˙$ which slowly loses O_2. Then RCO˙ abstracts H from the peracid to give $RCO_3^˙$ (then the chain mechanism can proceed) and RCHO which reacts rapidly with peracids in order to form two molecules of acid. We are now working on these two alternative mechanisms.

8 Conclusions

Peracids give rise to two processes of radical reactions.

1. In the case of *nucleophilic* radicals, mainly alkyl and cycloalkyl radicals, an S_H2 reaction occurs at the O−O peracid bond and results in an alcohol. If the

orbital containing the unpaired electron is not delocalized, the reactivity of the radical increases with its nucleophilicty, correlated with its ionization potential or more generally with the energy gap between its SOMO orbital and the σ^*_{O-O} peracid orbital. The consequences of this type of interaction are that a σ (having the unpaired electron in a quasi sp_3 orbital like the bicyclo[2,2,1]heptyl-1 or cyclopropyl radicals, or in a sp_2 orbital like the phenyl radical) is very unreactive and gives little or no alcohol, whereas a π radical (having the unpaired electron in a quasi p orbital like alkyl and cycloalkyl radicals) leads quantitatively to the alcohol. The reactive sequence for alkyl radicals is: tertiary > secondary > primary. This conclusion that reactivity varies with the radical nucleophilicity and not with its stability, contrary to the generally acknowledged axiom, has been observed in other cases in radical chemistry. If the radical is delocalized, the orbital overlap integral between the radical and the substrate becomes an important reactivity factor and a radical seemingly more nucleophilic on the basis of its ionization potential proves to be less reactive than a non-delocalized primary radical; this is the case for the benzyl radical.

The solvent may intervene by virtue of its H donor character; the radical coming from gives rise to an S_H2 reaction with the peracid to give solvent-alcohol (Scheme 15). The peracid, in this case, is an *hydroxylating* reagent.

2. If the radical is *electrophilic*, the reaction does not occur at the $O-O$ bond but, probably, there is addition onto the $C=O$ or abstraction of the H of the peroxyacid group. The peracid is decomposed to the acid by a mechanism not well understead and which is still under investigation.

9 References

1. Lefort D, Paquot C, Sorba J(1959) Bull Soc Chim France 1385
2. Lefort D, Sorba J, Rouillard D (1961) Bull Soc Chim France 2219
3. Lefort D, Sorba, J (1961) Bull Soc Chim France 2373
4. Vorobiev V, Lefort D, Sorba J, Rouillard D (1962) Bull Soc Chim France 1577
5. The Man L, Lefort D (1962) Bull Soc Chim France 827
6. Lefort D, Sorba J (1962) Oléagineux 17: 645
7. The Man L (1966) Bull Soc Chim France 652
8. Gruselle M, Fossey J, Lefort D (1970) Tetrah Letters 2069
9. Fossey J, Gruselle M, Lefort D (1970) Bull Soc Chim France 2635
10. Nedelec JY, Gruselle M Lefort D (1971) CR Acad Sc (C) 273: 1549
11. Nedelec JY, Lefort D (1972) Tetrah Letters 5073
12. Gruselle M, Tichy M, Lefort D (1972) Tetrahedron 28: 3885
13. Gruselle M, Lefort D (1973) Tetrahedron 29: 3035
14. Fossey J (1973) Tetrah Letters 1127
15. Nedelec JY, Lefort D (1975) Tetrahedron 31: 411
16. Gruselle M (1975) Tetrahedron 31: 2283
17. Gruselle M, Lefort D (1976) Tetrahedron 32: 2719
18. Gruselle M, Fossey J, Lefort D, Lamarre C, Richer JC (1976) Can J Chem 54: 905
19. Gruselle M, Nedelec JY (1978) Tetrahedron 34: 1813
20. Sorba J, Fossey J, Nedelec JY, Lefort D (1979) Tetrahedron 35: 1509
21. Fossey J, Lefort D (1980) Tetrahedron 36: 1023
22. Sorba J, Fossey J, Lefort D, Nedelec JY (1981) Tetrahedron 37: 69
23. Bégué JP, Lefort D, Truong Dinh T (1981) J Chem Soc Chem Commun 1086

24. Sorba J, Fossey J, Nedelec JY, Lefort D (1982) Tetrahedron 38: 2083
25. Lefort D, Nedelec JY (1982) Tetrahedron 38: 2681
26. Nedelec JY, Fossey J, Lefort D, Sorba J (1984) Can J Chem 62: 2317
27. Fossey J, Lefort D, Massoudi M, Nedelec JY, Sorba J (1985) Can J Chem 63: 678
28. Lefort D, Fossey J, Gruselle M, Nedelec, JY, Sorba J (1985) Tetrahedron 41: 4237
29. Fossey J, Lefort D, Massoudi M, Nedelec JY, Sorba J (1986) J Chem Soc Perkin Trans II 781
30. Fossey J, Lefort D, Sorba J (1986) J Org Chem 51: 3584
31. Fossey J, Grisel F, Lefort D, Sanderson W, Sorba J (1992) (in press)
32. Swain H, Silbert L, Miller J (1964) J Amer Chem Soc 86: 2562
33. Maillard B, Ingold K, Scaiano, J (1983) J Amer Chem Soc 105: 5085
34. Tiecco M, Testaferri L (1983) Abramovitch R (ed) In Reactive intermediates vol 3 Plenum Press, New York
35. Cherest M, Felkin H, Prudent N (1968) Tetrah Letters 2199
36. Cherest M, Felkin H (1968) Tetrah Letters 2205
37. Gloux J, Gugliemi M, Lemaire, H (1970) Mol Phys 19: 633
38. Kawamura T, Koyama T, Yonezawa T (1973) J Amer Chem Soc 95: 3220
39. Müller D, Perlberger J (1976) Helv Chim Acta 59: 2335
40. Kaptein R (1975) Advances in free radical chemisty 5: 319
41. Chateauneuf, J, Lusztyk J, Ingold K (1988) J Amer Chem Soc 110: 2886
42. Huang X, Dannenberg J (1991) J Org Chem 56: 5421

Cleaning-up Oxidations with Hydrogen Peroxide

K. M. Dear

Interox, Research & Development, Widnes Laboratory, P.O. Box 51, Moorfield Road, Widnes, Cheshire WA8 OFE, UK

Table of Contents

Whilst effluent treatment with H_2O_2 is possible in many cases to give environmentally more acceptable processes it is always preferable to improve the situation at source. H_2O_2 can be used in a range of applications to replace, reduce or handle more effectively environmentally unacceptable products. In performing these improvements the only byproducts are water and oxygen.

Topics in Current Chemistry, Vol. 164
© Springer-Verlag Berlin Heidelberg 1993

1 Introduction

The properties and reactivity of Hydrogen Peroxide make it an excellent candidate for improving the environmental acceptability of a wide range of chemical processes. This is primarily based upon the fact that its constituents are oxygen and water, the latter being the by-product of H_2O_2 reactions. In addition H_2O_2, and its derivatives, have a number of advantages.

1. Purity – H_2O_2 is a high purity reagent which is prepared in efficient high volume processes throughout the world.
2. Stability – When handled correctly H_2O_2 is a very stable compound over a range of conditions, typically losing less than 1% by weight of its active oxygen content per year.
3. Oxidation efficiency – H_2O_2 has a high active oxygen content due to its low molecular weight which makes it an efficient oxidant compared to many other oxidants such as transition metal based compounds.
4. Versatility – H_2O_2 can be used in both aqueous and organic media, under generally mild conditions and often using low excesses of the reagent.

H_2O_2 and its derivatives can be used for a range of applications within a number of industries for environmental benefit. H_2O_2 is used in the textile and paper industries for bleaching in direct competition to chlorine oxidants. Peroxygen reagents are good disinfectants in a number of industries, including the treatment of municipal effluents leading directly to cleaner beaches and inland waterways. The new generation of "green" household bleaching products contain peroxygen based active ingredients. There is a large potential for the use of H_2O_2 and its derivatives in the treatment of a range of industrial effluents. In the chemical processing industry H_2O_2 Chemistry offers environmental improvements in both replacing less attractive oxidants and opening up cleaner routes to desired products.

The bulk of this article will focus on this latter area of obtaining environmental benefits within the chemical processing industry. The main focus will be on used of peroxygen technology in the processes themselves with a short overview of the potential in effluent treatment.

2 Chlorine Replacement

Chorine based oxidants (particularly the element itself and sodium hypochlorite) are used in a number of areas of the chemical industry. Chlorinated intermediates such as epichlorhydrin and chlorinated solvents have traditionally been extensively used in a variety of chemical processes. Environmental pressure, both "consumer" based and legislative, is currently being exerted, and will increase in the future, for the reduction, and probably eventual removal, of the use of these reagents and intermediates. H_2O_2 based technology is well placed to offer environmentally acceptable alternatives and is already used in many areas for this reason.

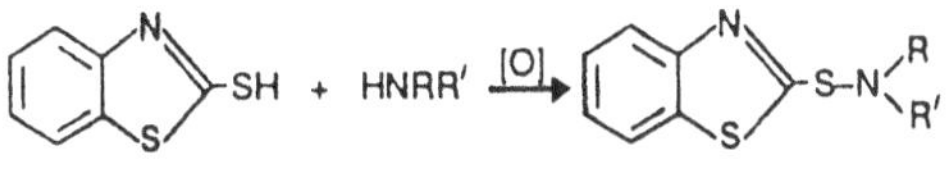

R	R′	Sulphenamide
H	C_6H_{11}	NCBS
$-(CH_2)_2O(CH_2)_2-$		NMBS
H	tBu	NTBBS

Fig. 1. Preparation of sulphenamides

A good example of the competitiveness between chlorine and peroxygen oxidants is the preparation of rubber accelerators such as sulphenamides. Sulphenamides are prepared by the oxidative coupling of a mercaptan with an amine (Fig. 1).

Sulphenamides can be prepared using a wide variety of oxidising agents. However, commercially, sodium hypochlorite or chlorine are most often used. With the increasing environmental pressure on chlorine oxidants, the industry is looking for alternative clean oxidants.

In addition to their inherent environmental unsuitability, chlorine based oxidants have a number of disadvantages.

1. Chlorinated by-products (principally chloramines) can be formed.
2. The process effluent has a high salt load.
3. The product requires several washing stages to reduce the salt content.

However, they do give very good chemical yields (95% for NCBS).

H_2O_2 circumvents all these problems and in addition gives better quality products (better sulphenamide stability). The main disadvantage at present when compared to chlorine oxidants is a lower yield (typically 77% for NCBS).

In the past this has meant that the chlorine oxidant process was significantly cheaper but increasing effluent treatment costs and environmentally based legislation is altering this balance in favour of H_2O_2. This is an aera of active research for Interox which will result in a further redressing of the balance in H_2O_2's favour.

It has previously been shown that a compromise of the disadvantage of the two systems can be obtained by combining them (Fig. 2). Thus the yield can be improved over H_2O_2 alone whilst reducing the environmental disadvantages of NaOCl alone.

The use of chlorinated intermediates in processes often gives products which contain chlorinated impurities that are detrimental to the products' performance or toxicology.

An example of this is in the preparation of glycidyl compounds.

These types of compounds (see Fig. 3 for examples) are important in the polymer and resin area. They are traditionally prepared by the condensation of an appropriate precursor (e.g. a phenol) with epichlorhydrin (Fig. 4). Whilst this process works well, the product can react further to give homologues and relatively high levels or organic chlorine are retained within the final product. These by-products give problems with both the performance and the toxicology of the products.

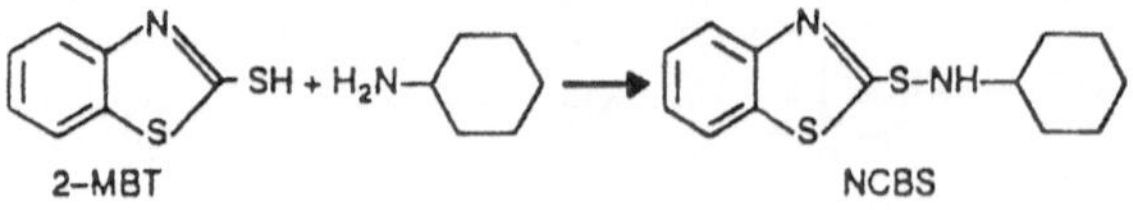

K. M. Dear

2–MBT

NCBS

Conditions: 2MBT: Amine 1:3
20% excess oxidant
50°C, pH 10–10.5

Results:

$NaOCl:H_2O_2$	Yield (%)
0:1	77
1:3	82
1:0	95

Fig. 2. Preparation of NCBS with mixed oxidants

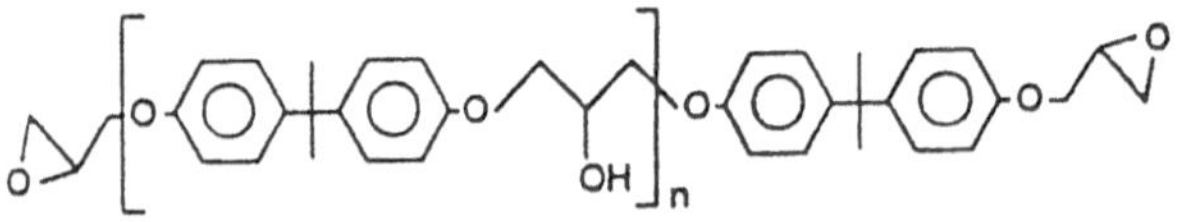

Bisphenol A based diglycidyl ethers

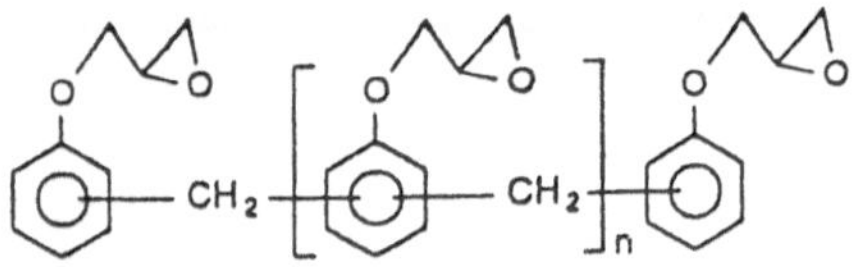

Novolacs

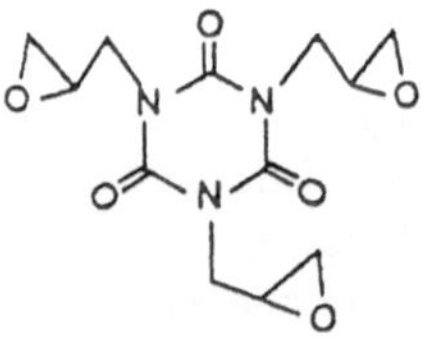

Heterocyclic glycidyl imides and amides

Fig. 3. Some examples of commercially used glycidyl compound types

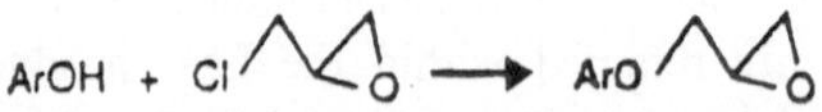

Fig. 4. Preparation of aryl glycidyl compounds via epichlorhydrin

118

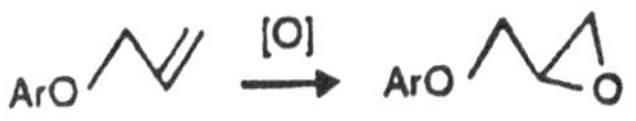

Fig. 5. Preparation of aryl glycidyl compounds by a peroxygen route

H_2O_2 technology offers the potential to circumvent these problems via an alternative route (Fig. 5).

The epoxidation of an appropriate O-allyl group gives the appropriate glycidyl compounds without the detrimental by-products of the epichlorhydrin route. However this transformation is not an easy one and whilst it works well for mono-allyl compounds, polyallyl substrates offer more problems.

A monoallyl substrate such as phenyl allyl ether can be epoxidised with Paynes reagent or peracids in good yield (Fig. 6).

$$PhOCH_2CH=CH_2 \quad \xrightarrow{PhCN/H_2O_2} \quad PhOCH_2\overset{O}{\overset{\diagup\diagdown}{CH-CH_2}}$$

Conversion = 72%

Selectivity = 63%

Fig. 6. Epoxidation of phenyl allyl ether with Paynes reagent

Although the polyglycidyl compounds are more·difficult to produce, Interox has developed a peracid based method and a metal catalysed H_2O_2 method which both give good results.

3 Replacement of Transition Metal Oxidants

Transition metal oxidants such as manganese and chromium oxidants have been widely used in the chemical industry over the years. They have a major disadvantage in that they produce large volumes of effluent containing the transition metals which are subject to more and more strenuous controls on discharge levels. Supported reagents or effluent recycle could be considered but neither is easy on an industrial scale and they are cures rather than prevention of the problem.

In many cases, peroxygen technology can be used to avoid the use of the transition metal altogether. Transition metal oxidants are traditionally used for alcohol or aldehyde oxidation and side chain oxidation of aromatic compounds such as toluenes.

The oxidation of alcohols to corresponding carbonyl compounds is readily achieved with H_2O_2 and a molybdenum catalyst (Fig. 7) [1].

The reaction is performed under phase transfer conditions and proceeds to the aldehyde without overreaction to the acid in the case of the benzyl alcohol.

CH₂OH → 70% → CHO

H₂O₂/ (NH₄)₆Mo₇O₂₄

→ 92% →

Fig. 7. Oxidation of alcohols with molybdenum catalysed H_2O_2

Tungsten catalysts can also be used giving excellent yields of aldehydes and ketones (Fig. 8) [2].

Product	Yield (%)
	96
	97
PhCHO	85

Fig. 8. Oxidation of alcohols with H_2O_2 and W(VI) catalysts

Although a transition metal is still used here it is in catalytic amounts (about 5% w/w) and immobilisation of catalysts is an inherently simpler problem than immobilising stoichiometric oxidants.

Aldehyde to acid oxidations are another class of transformations which are often carried out using transition metal reagents. A project that recently arose in this area for Interox concerned the oxidation of cumene aldehyde to the acid (Fig. 9).

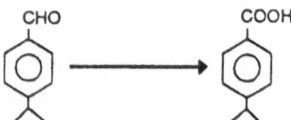

CHO → COOH

Fig. 9. Oxidation of cumene aldehyde

The reaction can be carried out using $KMnO_4$ but yields are moderate and an impurity where the isopropyl groups was hydroxylated was produced.

When oxidising benzaldehydes with peroxygen compounds one must be aware of the possible Dakin reaction which can occur as a side reaction. As one would expect alkaline conditions to promote the Dakin reaction we first studied peracid systems (Fig. 10). Surprisingly, significant levels of phenol were discovered in this reaction. However, when alkaline H_2O_2 was used excellent yields of acid were obtained with no phenol detected and no hydroxylation of the isopropyl group.

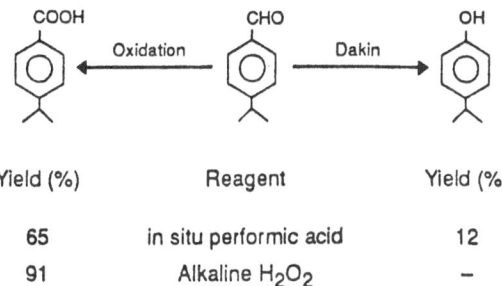

Yield (%)	Reagent	Yield (%)
65	in situ performic acid	12
91	Alkaline H₂O₂	–

Fig. 10. Benzaldehyde oxidation

A similar result with alkaline H_2O_2 had been reported [3]. However, the Interox method showed significant improvement by use of stabiliser packages for the H_2O_2 and control of pH (Table 1).

Table 1. Oxidation of cumene aldehyde with alkaline H_2O_2

Additional Stabiliser	pH	H₂O₂ used (mole eq.)	Yield (%)
None	12	5.1	85
Dequest 2060 [4]	12	4.4	88
Mykon CIX [5]	12	3.9	90

The oxidation of a hydrocarbon side chain attached to an aromatic ring is traditionally a difficult transformation to accomplish. The most common transformation is the conversion of toluenes to benzaldehydes and benzoic acids (Fig. 11). The most difficult stage is the introduction of oxygen as subsequent interconversion between oxidation levels is relatively easy as we have already seen.

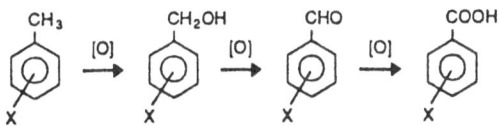

Fig. 11. Side chain oxidation of toluenes

The oxidation of the toluence can be achieved directly by transition metal oxidants [6]. However, not only do they have their effluent based environment problems, they also tend to be non-selective to the range of products produced.

Interox has developed an indirect approach to this range of products based on a photochemical bromination followed by hydrolysis and oxidation steps (Fig. 12). The bromine a potentially environmentally unfriendly reagent in its own right, is generated in situ from H_2O_2 and HBr.

The selectivity of the bromination can be controlled by the reaction conditions to give either mono or dibromo products [7]. These can be hydrolysed and oxidised to give a required product selectivity.

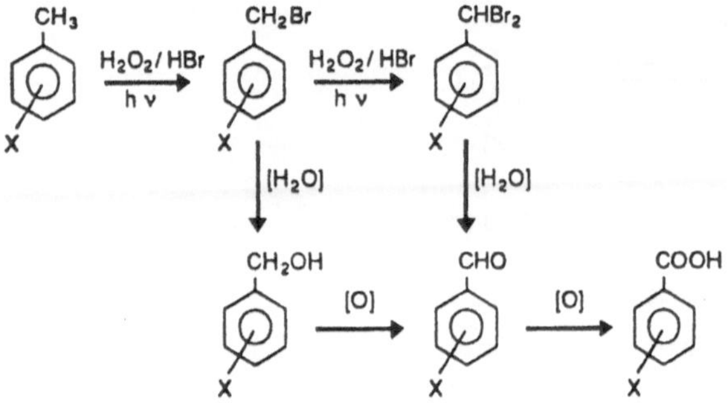

Fig. 12. Interox technology for side chain oxidation

The technology works best for deactivated substrates but can be adapted to give a wide range of products (Fig. 13).

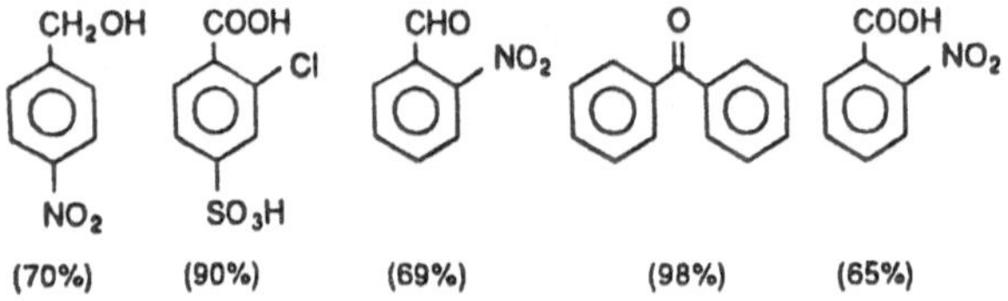

Fig. 13. Products available via Interox side chain oxidation technology (yields from hydrocarbon substrate in brackets)

The yields of these products are generally very good despite the multi-stage processing. In some examples the whole process takes place in one step when the brominated products are hydrolysed by aqueous HBr and the subsequent oxidations proceed by bromine. The production of benzophenones is a good example of this [8].

4 Halogen Handling

Halogens, principally chlorine and bromine, are widely used in the chemical industry. There are a number of concerns regarding the saftey and environmental aspects of handling such compounds. H_2O_2 provides a method of generating the halogens in situ, as we have seen earlier, and thus avoiding the potential environmental problems in handling the elements directly. Moreover, H_2O_2 can also be used to recover the halogen from the waste effluent streams to reduce halide/halogen discharge levels.

The use of H_2O_2/HX in place of the parent halogen gives a number of environmental advantages including
1. Full utilisation of the halogen (low halide discharge levels from processes).
2. The major by-product of the process is water.
3. The halogen is generated in situ avoiding the environmental hazards of bulk storage.

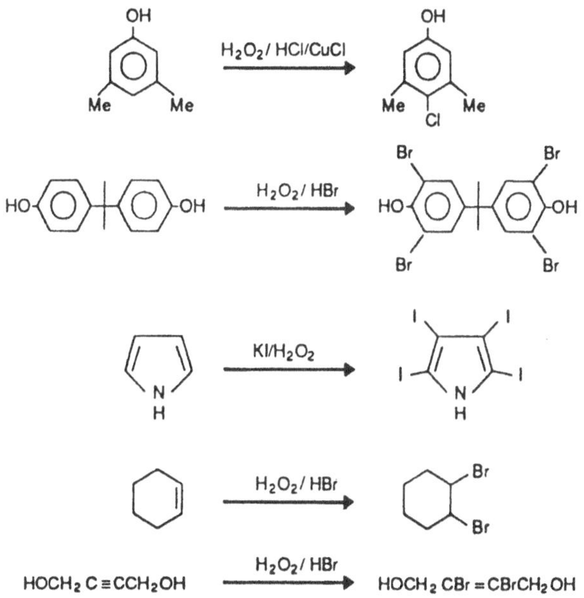

Fig. 14.
Halogenation with H_2O_2

The major disadvantage of the system is that it is two phase, aqueous/organic. This limits its use in a few cases (e.g. water sensitive materials) but generally is not a problematical drawback in practice.

The system can be used for halogenation of both aliphatic and aromatic substrates (Fig. 14).

The reaction between H_2O_2 and HBr can also be used to reclaim bromine from a bromide effluent (Fig. 15).

$$H_2O_2 + 2\,HBr \longrightarrow Br_2\uparrow + 2\,H_2O$$

Fig. 15. Liberation of bromine from HBr solutions with H_2O_2

Interox has shown that the reaction is very fast (being complete in 30–40 seconds), highly exothermic and that $>95\%$ yield of bromine can be obtained. The use of acid catalysis will improve the reaction enabling all strengths of HBr solution to be utilised.

Although H_2O_2 will react with bromine to give, ultimately, oxygen and HBr [9], this decomposition reaction is very slow compared to the formation of bromine and has little effect on process economics. Interox has also shown that a plant to recover 200 tonnes per annum of bromine would not have a significant cost when compared to the operating plant and would have a payback period of only 1 to 2 years.

The competitive technology is based on chlorine which although much cheaper on raw material costs, is very little cheaper when operating costs are considered.

The main advantage of the H_2O_2 method is that an environmental problem is cured without the use of an environmentally unacceptable agent.

5 Product Quality

In some reactions the use of peroxygen reagents will give products of better quality than the competitive reagents (cf. sulphenamide preparation earlier). In some instances this improvement in quality will take the form of reducing the level of an environmentally unwanted by-product. This is the case in nitrosamine formation in amine oxide manufacture. Recent work in our American laboratories showed that the use of H_2O_2 in amine oxide formation can significantly reduce nitrosamine levels in the product and and that certain stabiliser packages can make further radical improvements (Table 2).

Table 2. Reduction of nitrosamines amine oxide production

$$C_{12}H_{25}(CH_3)_2N \xrightarrow{H_2O_2} C_{12}H_{25}(CH_3)_2N^+ - O^-$$

H_2O_2 Grade	Conversion (%)	Yield (%)	Nitrosoamine level (ppb)
Commercial Amine oxide sample	—	—	3900
USA Storage	87	86	1260
Dequest 2066 Stabilised (4)	97	95	250
Mykon CIX (5)	97	96	250

The table shows that, in dodecyldimethylamine oxide production, the use of USA storage grade H_2O_2 reduces nitrosamines by a factor of 3. Addition of a phosphonate based H_2O_2 stabilisation package reduces this level by a factor of 5 whilst increasing yield and conversion.

Similar effects can be noted in other reactions when special grades of H_2O_2 are used that have been tailored to that reaction.

6 Effluent Treatment

When noxious and environmentally hazardous effluents are products from processes, H_2O_2 can often provide a clean method of rendering the effluent environmentally more acceptable. Some examples are as follows:

1. Sulphur − Effluents which contain noxious sulphur compounds (e.g. H_2S, sulphides and mercaptans) can be treated with H_2O_2. H_2S and S^{2-} can be treated at high pH to give the sulphate ion and water. Organic mercaptans can be oxidised with H_2O_2 using a tungsten catalyst to give sulphonic acids.
2. Phenols − Phenol containing effluents can be efficiently destroyed by using Fenton's reagent (H_2O_2 with iron catalysis). This reagent produces the highly reactive OH˙ radical.
3. Cyanide − The use of H_2O_2 for cyanide detoxification is increasing. Many effluents are treated with H_2O_2 at pH 9 to give cyanate anion which is subsequently hydrolysed to carbonic acid. Some effluents, which contain

complex cyanides e.g. in the gold mining industry, can be more effectively treated with Caro's acid (H_2SO_5) and Interox has recently developed new technology to produce the reagent which makes it highly competitive.

4. Gas scrubbing — Effluent gases containing low molecular weight mercaptans and amines can be scrubbed with H_2O_2 systems. Methods using copper or tungsten catalysts can be used to give sulphonates, hydroxylamines, oximes or amine oxides depending on the effluent. Sulphur dioxides can be similarly absorbed and treated as can NO_x gases as long as good gas-liquid contact is maintained. Other alkaline or hypochlorite based systems have been shown to be less effective. A recent variation of NO_x treatment has been developed in the stainless steel industry. Addition of H_2O_2 to the HNO_3 pickling bath completely suppresses NO_x emissions, improves process economics and can produce quality benefits.

5. Organic phosphate — Recent work between BNFL and Interox (11) has resulted in a chromium catalysed H_2O_2 method for destroying tributylphosphate in the nuclear industry. CO_2, H_2O and inorganic phosphate are the only products.

7 References

1. Trost BM, Masuyama Y (1984) Tetrahedron Lett 25: 173
2. Jacobson SE, Muccigrosso DA, Maro F (1979) J Org Chem 44: 921;
   Bortolini O, Conte V, Di Furia F, Modena G (1985) Nouv J Chim 9: 147; (1986) J Org Chem 51: 2661
3. WAKO Pure Chem Ind (1988) Japanese Pat J 63264551-A
4. Dequest is a Monsanto Trade Name
5. Mykon is a Warwick International Trade Name. Mykon CIX is available solely from Interox
6. See inter alia: Cooper TA, Waters WA, (1967) J Chem Soc (B): 687; Heiba EI, Dessau RM, Koehl WJ (1969) J Am Chem Soc 91: 138, 6830; Gilmore JR, Mellor JM (1970) J Chem Soc Chem Commun: 507
7. Interox Chemicals Ltd (1989) EP 336568, (1989) EP 336567 and (1990) US Pat 4943358
8. Interox Chemicals Ltd. (1989) EP 334511
9. See inter alia: Livingstone R, Schoeld EA (1936) J Am Chem Soc 58: 1244
10. Interox Chemicals Ltd. (1991) EP 409043
11. Interox Chemicals Ltd/BNFL (1990) US Pat 4950425

Author Index Volumes 151–164

The volume numbers are printed in italics

Kavarnos, G. J.: Fundamental Concepts of Photoinduced Electron Transfer. *156*, 21–58 (1990).

Khairutdinov, R. F., see Zamaraev, K. I.: *163*, 1–94 (1992).

Kim, J. I., Stumpe, R., and Klenze, R.: Laser-induced Photoacoustic Spectroscopy for the Speciation of Transuranic Elements in Natural Aquatic Systems. *157*, 129–180 (1990).

Klaffke, W. see Thiem, J.: *154*, 285–332 (1990).

Klein, D. J.: Semiempirical Valence Bond Views for Benzenoid Hydrocarbons. *153*, 57–84 (1990).

Klein, D. J., see Chen, R. S.: *153*, 227–254 (1990).

Klenze, R., see Kim, J. I.: *157*, 129–180 (1990).

Knops, P., Sendhoff, N., Mekelburger, H.-B., Vögtle, F.: High Dilution Reactions — New Synthetic Applications *161*, 1–36 (1991).

Koepp, E., see Ostrowicky, A.: *161*, 37–68 (1991).

Kostikov, R. R., Molchanov, A. P., and Hopf, H.: Gem-Dihalocyclopropanos in Organic Synthesis. *155*, 41–80 (1990).

Krogh, E., and Wan, P.: Photoinduced Electron Transfer of Carbanions and Carbacations. *156*, 93–116 (1990).

Kunkeley, H., see Vogler, A.: *158*, 1–30 (1990).

Kuwajima, I. and Nakamura, E.: Metal Homoenolates from Siloxycyclopropanes. *155*, 1–39 (1990).

Lange, F., see Mandelkow, E.: *151*, 9–29 (1989).

Lefort, D., see Fossey, J.: *164*, 99–113 (1993).

Lopez, L.: Photoinduced Electron Transfer Oxygenations. *156*, 117–166 (1990).

Lymar, S. V., Parmon, V. N., and Zamarev, K. I.: Photoinduced Electron Transfer Across Membranes. *159*, 1–66 (1991).

Mandelkow, E., Lange, G., Mandelkow, E.-M.: Applications of Synchrotron Radiation to the Study of Biopolymers in Solution: Time-Resolved X-Ray Scattering of Microtubule Self-Assembly and Oscillations. *151*, 9–29 (1989).

Mandelkow, E.-M., see Mandelkow, E.: *151*, 9–29 (1989).

Mattay, J., and Vondenhof, M.: Contact and Solvent-Separated Radical Ion Pairs in Organic Photochemistry. *159*, 219–255 (1991).

Mekelburger, H.-B., see Knops, P.: *161*, 1–36 (1991).

Meng, Q., Hesse, M.: Ring Closure Methods in the Synthesis of Macrocyclic Natural Products *161*, 107–176 (1991).

Merz, A.: Chemically Modified Electrodes. *152*, 49–90 (1989).

Meyer, B.: Conformational Aspects of Oligosaccharides. *154*, 141–208 (1990).

Moffat, J. K., Helliwell, J.: The Laue Method and its Use in Time-Resolved Crystallography. *151*, 61–74 (1989).

Molchanov, A. P., see Kostikov, R. R.: *155*, 41–80 (1990).

Moore, T. A., see Gust, D.: *159*, 103–152 (1991).

Nakamura, E., see Kuwajima, I.: *155*, 1–39 (1990).

Okuda, J.: Transition Metal Complexes of Sterically Demanding Cyclopentadienyl Ligands. *160*, 97–146 (1991).

Ostrowicky, A., Koepp, E., Vögtle, F.: The "Vesium Effect": Syntheses of Medio- and Macrocyclic Compounds *161*, 37–68 (1991).

Parmon, V. N., see Lymar, S. V.: *159*, 1–66 (1991).

Polian, A., see Fontaine, A.: *151*, 179–203 (1989).

Raimondi, M., see Copper, D. L.: *153*, 41–56 (1990).

Riekel, C.: Experimental Possibilities in Small Angle Scattering at the European Synchrotron Radiation Facility. *151*, 205–229 (1989).